# Master Electrician Exam Prep:

Disclaimer

This publication is intended for informational purposes only and does not constitute legal, professional, or any other kind of advice. The author and publisher have made every effort to ensure the accuracy and completeness of the information provided herein. However, they cannot guarantee the applicability of the content to any specific situation or circumstance, nor can they warrant that the information will remain current or error-free.

The author and publisher disclaim any responsibility for errors, inaccuracies, or omissions in this publication, and they shall not be held liable for any direct, indirect, incidental, special, consequential, or other damages resulting from the use of, or reliance on, the information provided in this book. Readers are advised to consult with a qualified professional before making any decisions or taking any actions based on the content in this publication.

This book is not affiliated with, endorsed by, or sponsored by any organization, agency, or authority responsible for the administration or regulation of electrical systems, including but not limited to the National Electrical Code (NEC) or any state or local agencies. It is the reader's responsibility to comply with all applicable codes, regulations, and requirements when working with electrical systems.

*PAGE 2 Introduction.*
*PAGE 6 Electrical Theory.*
*PAGE 16 National Electrical Code (NEC).*
*PAGE 26 Safety.*
*PAGE 33 Blueprint Reading and Project Management.*
*PAGE 43 Electrical Calculations.*
*PAGE 54 Installation and Troubleshooting.*
*PAGE 66 - 215 Practice Exam Section with 500+ Practice Exam Questions.*

INTRODUCTION:
Welcome to the Ultimate Master Electrician Exam Study Guide! We understand that pursuing your dreams can be challenging, and the path to becoming a master electrician is no exception. That's why we've created this comprehensive resource to support you every step of the way, making your journey smoother and your success well within reach.

This study guide is designed to be your trusted companion as you prepare for the Master Electrician exam, covering everything you need to know in a clear, concise, and accessible manner. We've left no stone unturned, ensuring that you're equipped with the knowledge and skills necessary to confidently navigate the complexities of electrical theory, the National Electrical Code (NEC), safety, blueprint reading, project management, electrical calculations, installation, and troubleshooting.

Our goal is to not only help you pass the exam but to also instill a sense of pride and confidence in your abilities as an electrician. We understand that setbacks can be disheartening, and sometimes even the smallest of failures can cast a shadow over your dreams. But rest assured, this guide is here to justify your hard work and dedication, offering an all-in-one resource that will help you transform those challenges into opportunities for growth and success.

We know that the prospect of the exam can be daunting, but we are here to allay your fears and provide you with the tools and strategies necessary to excel. In addition to the thorough coverage of essential topics, this guide also includes practice exams and review questions to reinforce your learning and build your test-taking confidence.

As you embark on this journey, remember that you're not alone. We are here to support you, offering encouragement, guidance, and expert knowledge to help you achieve your dreams. We believe in your potential, and we're confident that with determination, persistence, and the help of this study guide, you'll soon join the ranks of successful master electricians.

Now, let's begin this exciting adventure together. It's time to turn the page, dive into the content, and take the first step toward a brighter, more electrifying future!

Wishing you the best of luck.

**Purpose of the Study Guide**

Embarking on the journey to become a master electrician is a remarkable and ambitious endeavor. With this study guide, our mission is to provide you with a comprehensive, all-encompassing resource that equips you with the knowledge and skills required to excel in the Master Electrician exam and, ultimately, in your professional career.

We've meticulously crafted this guide to be your one-stop-shop for all your preparation needs. Each chapter delves into essential topics, breaking them down into easily digestible sections that cater to various learning styles. From the intricacies of electrical theory to the complexities of the

National Electrical Code, safety regulations, and troubleshooting techniques, this study guide aims to cover all the bases, leaving no stone unturned.

Our objective is to empower you with the confidence and expertise needed to conquer the exam and thrive in the electrical industry. To do so, we've included practice exams and review questions throughout the guide, allowing you to assess your understanding, reinforce your learning, and develop effective test-taking strategies.

In crafting this study guide, we've prioritized clarity, practicality, and an engaging tone to create a learning experience that's both enjoyable and effective. Our aim is to make your preparation process as smooth and efficient as possible, enabling you to focus on honing your skills and expanding your knowledge.

In essence, the purpose of this study guide is to serve as your steadfast companion on the road to becoming a master electrician, ensuring that you have all the tools and resources necessary to achieve your goals and fulfill your dreams. With dedication, determination, and the support of this guide, we're confident that you'll be well on your way to a successful and rewarding career in the electrical field.

**Overview of the Master Electrician Exam**

The Master Electrician exam is a crucial milestone for individuals aspiring to become licensed master electricians. This rigorous test evaluates your comprehension of essential concepts, principles, and practices relevant to the electrical industry. By passing this exam, you demonstrate your proficiency and commitment to upholding the highest standards of safety, quality, and professionalism.

The exam typically consists of a series of multiple-choice questions designed to assess your understanding of a wide range of topics, including electrical theory, the National Electrical Code (NEC), safety regulations, blueprint reading, project management, electrical calculations, installation, and troubleshooting. The exact content and structure of the exam may vary depending on your jurisdiction, but the primary goal remains the same: to ascertain your competence as a master electrician.

Before attempting the exam, it's essential to meet specific eligibility requirements, which often include a combination of education, work experience, and the completion of a journeyman electrician license. These prerequisites help ensure that candidates possess the foundational knowledge and practical skills necessary for success at the master electrician level.

The exam itself is typically administered under strict time constraints, requiring candidates to demonstrate not only their knowledge but also their ability to think critically and efficiently under pressure. This aspect of the test highlights the importance of developing effective test-taking strategies and honing your time management skills.

To adequately prepare for the Master Electrician exam, it's crucial to develop a comprehensive understanding of the various topics covered, as well as the nuances of the NEC and other relevant codes and regulations. By thoroughly studying the material and practicing with sample questions, you'll build the confidence and expertise necessary to tackle the exam head-on and achieve your goal of becoming a licensed master electrician.

In summary, the Master Electrician exam is a rigorous and comprehensive evaluation of your knowledge, skills, and professionalism within the electrical field. By dedicating yourself to thorough preparation and leveraging the support of this study guide, you'll be well-equipped to conquer the exam and embark on a fulfilling and successful career as a master electrician.

**Test-Taking Strategies and Tips**
To maximize your chances of success on the Master Electrician exam, it's essential to adopt effective test-taking strategies and techniques. Here are some tips to help you approach the exam with confidence and make the most of your preparation:

1. Develop a study plan: Create a structured study schedule that allocates ample time to review each topic and practice exam questions. Ensure your plan includes breaks and periodic review sessions to reinforce your learning and maintain focus.
2. Understand the exam format: Familiarize yourself with the structure and content of the exam, including the types of questions, time constraints, and scoring system. Knowing what to expect can help you better allocate your time and effort during the test.
3. Practice time management: Develop your time management skills by setting time limits when practicing sample questions or taking mock exams. This will help you gauge your pacing and ensure you have enough time to answer all questions on the actual exam.
4. Read questions carefully: Thoroughly read each question and all answer choices before selecting an answer. Misreading a question or misunderstanding the options can lead to incorrect answers, even if you know the material well.
5. Eliminate unlikely answers: If you're unsure of the correct answer, eliminate options that seem implausible or irrelevant. This will increase your chances of selecting the correct answer, even if you have to make an educated guess.
6. Answer every question: Since most Master Electrician exams consist of multiple-choice questions, it's in your best interest to answer all questions, even if you're uncertain. There is typically no penalty for guessing, so don't leave any questions unanswered.
7. Use the process of elimination: When faced with a difficult question, try working through the problem by eliminating incorrect answers one by one. This can help you narrow down the choices and arrive at the correct answer more easily.
8. Stay calm and focused: Stress and anxiety can negatively impact your performance. Practice relaxation techniques, such as deep breathing or visualization, to stay calm and maintain concentration during the exam.
9. Review your answers: If you have time remaining after completing the exam, review your answers to ensure you haven't made any careless mistakes or misread any questions. Be cautious when changing answers, as your first instinct is often correct.
10. Prepare mentally and physically: Get plenty of rest, eat well, and exercise in the days leading up to the exam. Being in good physical and mental health will help you perform at your best.

By incorporating these test-taking strategies and tips into your preparation, you'll be better equipped to approach the Master Electrician exam with confidence and poise, increasing your chances of success and paving the way for a thriving career in the electrical field.

# **Electrical Theory**

Welcome to the Electrical Theory chapter! As an aspiring master electrician, a solid understanding of electrical theory is crucial for your success, both on the exam and in your professional career. This chapter aims to provide you with a thorough and engaging exploration of the fundamental principles and concepts that underpin the world of electricity.

Electrical theory encompasses a wide array of topics, ranging from basic principles such as Ohm's Law and Kirchhoff's Laws to more advanced concepts like alternating current (AC), power factor, and transformers. By delving into these essential concepts, you'll develop the foundational knowledge necessary to effectively design, install, troubleshoot, and maintain electrical systems.

Throughout this chapter, we will break down complex topics into easily digestible sections, using clear explanations, practical examples, and real-world applications to enhance your understanding. You'll learn about the behavior of electrical circuits, the properties of various electrical components, and the key calculations needed to analyze and solve electrical problems.

In addition to expanding your theoretical knowledge, this chapter will also help you build the practical skills required for success in the electrical field. By understanding the principles that govern electrical systems, you'll be better equipped to navigate the challenges you'll face as a master electrician, from interpreting blueprints and adhering to code requirements to diagnosing and rectifying faults.

As you progress through this chapter, we encourage you to take your time, work through the examples, and engage with the material at your own pace. Remember, electrical theory is the bedrock upon which your career as a master electrician is built, so invest the time and effort necessary to truly grasp the concepts presented.

Let's embark on this exciting journey together and dive into the fascinating world of electrical theory. Your path to becoming a master electrician starts here!

Voltage, current, resistance, and power are fundamental concepts in the study of electricity. Understanding these principles is essential for mastering electrical theory and becoming a skilled electrician. Let's dive into each of these concepts and explore their significance in the world of electricity.

1. Voltage (V): Voltage, also known as electric potential difference, is the force that drives the flow of electrons (electric charge) through a conductor. It is measured in volts (V) and can be thought of as the "pressure" that pushes electric charge through a circuit. Voltage can be produced by various sources such as batteries, generators, and solar panels. In a circuit, the voltage across a component represents the energy provided or consumed per unit of electric charge passing through it.
2. Current (I): Current is the flow of electric charge through a conductor, such as a wire or a circuit component. It is measured in amperes (A) and is directly proportional to the

voltage applied across the conductor and inversely proportional to the resistance of the conductor. Current can flow in two directions: direct current (DC), where the flow is constant and unidirectional, and alternating current (AC), where the flow periodically changes direction.
3. Resistance (R): Resistance is a property of a material that opposes the flow of electric current. It is measured in ohms (Ω) and depends on factors such as the material's resistivity, length, cross-sectional area, and temperature. Conductive materials, like copper and aluminum, have low resistance, while insulative materials, such as rubber and plastic, have high resistance. In a circuit, resistance determines how much voltage is required to drive a specific current through a component or conductor.
4. Power (P): Power is the rate at which electrical energy is converted into other forms of energy, such as heat, light, or motion. It is measured in watts (W) and is calculated by multiplying voltage (V) and current (I): $P = V \times I$. In a circuit, power represents the amount of work done by electrical energy as it flows through components, such as resistors, capacitors, or inductive loads like motors. Understanding power consumption and efficiency is essential for designing and managing electrical systems.

In summary, voltage, current, resistance, and power are the foundational concepts in electrical theory. They describe the behavior of electrical circuits and the relationships between electrical quantities. By understanding these principles, you'll be better equipped to analyze, design, and troubleshoot electrical systems, setting the stage for success as a master electrician.

**Ohm's Law** is a fundamental principle in electrical theory, named after the German physicist Georg Simon Ohm. It establishes a relationship between voltage (V), current (I), and resistance (R) in an electrical circuit. The law is mathematically expressed as $V = I \times R$, which states that voltage across a conductor is equal to the product of the current flowing through it and its resistance.

The significance of Ohm's Law lies in its ability to simplify electrical calculations and help us understand the behavior of electrical circuits. It serves as a foundation for analyzing, designing, and troubleshooting electrical systems, making it a key concept for any master electrician. Practical applications of Ohm's Law are numerous and span various aspects of electrical work. Here are a few examples:
1. Circuit analysis: Ohm's Law allows electricians to calculate the voltage drop across individual components in a circuit, which is essential for determining the proper size and rating of components, such as wires, fuses, and switches.
2. Power consumption: By combining Ohm's Law with the power formula ($P = V \times I$), electricians can calculate the power consumed by a component, enabling them to make informed decisions about energy efficiency and component selection.
3. Troubleshooting: Ohm's Law can help electricians identify faulty components in a circuit. For example, if the voltage and current measurements do not align with the expected values based on the component's resistance, it may indicate a problem that requires further investigation.
4. Resistor sizing: Electricians can use Ohm's Law to determine the appropriate resistor value needed to limit the current flow in a circuit, ensuring the safe operation of connected components, such as LEDs or other sensitive electronics.

5. Voltage regulation: Ohm's Law plays a crucial role in designing voltage regulation circuits, such as voltage dividers, which are essential for providing stable and precise voltage levels to sensitive electronic devices.

In conclusion, Ohm's Law is a vital tool in the electrician's arsenal, enabling them to navigate the complexities of electrical systems with confidence. By understanding and applying this foundational principle, you'll be well-equipped to tackle a wide range of electrical challenges on your journey to becoming a master electrician.

**Kirchhoff's Laws**, named after the German physicist Gustav Kirchhoff, are two fundamental principles used to analyze electrical circuits. These laws, known as Kirchhoff's Voltage Law (KVL) and Kirchhoff's Current Law (KCL), help electricians understand the behavior of circuits and solve complex problems involving multiple components and connections.

1. Kirchhoff's Voltage Law (KVL): KVL states that the sum of the voltages around any closed loop in a circuit is equal to zero. This means that the total voltage supplied by sources in the loop is equal to the total voltage drop across all components in the same loop. KVL is a direct consequence of the conservation of energy principle, which asserts that energy cannot be created or destroyed.

KVL helps electricians analyze circuits by providing a systematic approach to determine the voltage across individual components. This is particularly useful when working with complex circuits, such as those with multiple voltage sources or parallel and series connections.

2. Kirchhoff's Current Law (KCL): KCL states that the sum of currents entering a junction (node) in a circuit is equal to the sum of currents leaving that junction. In simpler terms, the total current coming into a junction must equal the total current going out. KCL is based on the principle of conservation of charge, which implies that electric charge cannot be created or destroyed.

KCL is essential for electricians when analyzing circuits with multiple current paths, such as parallel circuits or circuits with multiple branches. It helps determine the distribution of current in various parts of the circuit, which is critical for selecting the correct components and ensuring the safe operation of the electrical system.

In summary, Kirchhoff's Laws play a crucial role in analyzing electrical circuits. They provide a systematic framework for understanding circuit behavior and solving complex problems. By mastering KVL and KCL, you will be better equipped to design, troubleshoot, and maintain electrical systems, paving the way for success as a master electrician.

**Series and parallel circuits** are two fundamental circuit configurations that have distinct characteristics and implications for voltage, current, and resistance. Understanding these differences is essential for analyzing, designing, and troubleshooting electrical systems.

1. Series Circuits: In a series circuit, components are connected end-to-end such that there is only one path for current to flow. The key characteristics of series circuits are:
   - Voltage: The total voltage across the circuit is divided among the components, with each component experiencing a voltage drop proportional to its resistance. The sum of these voltage drops equals the total voltage supplied by the source: $V\_total = V1 + V2 + ... + Vn$.

- Current: The current flowing through each component in a series circuit is the same, as there is only one path for current to flow. This means that the total current supplied by the source is equal to the current through each component: $I_{total} = I_1 = I_2 = ... = I_n$.
- Resistance: The total resistance in a series circuit is the sum of the individual resistances of the components: $R_{total} = R_1 + R_2 + ... + R_n$.
2. Parallel Circuits: In a parallel circuit, components are connected across common voltage points such that multiple current paths are created. The key characteristics of parallel circuits are:
- Voltage: The voltage across each component in a parallel circuit is the same, equal to the voltage supplied by the source: $V_{total} = V_1 = V_2 = ... = V_n$.
- Current: The total current supplied by the source is divided among the components, with each component drawing a current proportional to its resistance. The sum of these currents equals the total current supplied by the source: $I_{total} = I_1 + I_2 + ... + I_n$.
- Resistance: The total resistance in a parallel circuit can be found using the reciprocal formula: $1/R_{total} = 1/R_1 + 1/R_2 + ... + 1/R_n$. The total resistance in a parallel circuit is always less than the smallest individual resistance.

In summary, series circuits have a single current path, with voltage distributed among components and constant current throughout. In contrast, parallel circuits have multiple current paths, with constant voltage across components and current distributed among them. Understanding these differences and their implications for voltage, current, and resistance is essential for mastering electrical theory and becoming a skilled electrician.

**Alternating current** (AC) is an electrical current that periodically reverses its direction, unlike direct current (DC), which flows in a constant direction. AC is the most common form of electrical power used in residential, commercial, and industrial applications due to its numerous advantages over DC.

Advantages of AC over DC include:
1. Ease of transformation: AC voltage levels can be easily transformed up or down using transformers, which makes it more suitable for long-distance power transmission. High voltages are used for transmission to reduce power loss, and then transformed to lower voltages for safer use in homes and businesses.
2. Generation and distribution: AC is simpler to generate and distribute than DC, thanks to the widespread availability and efficiency of AC generators, such as alternators, and the extensive AC power grid infrastructure.
3. AC motors: AC motors are generally more efficient and require less maintenance than their DC counterparts. This makes AC-powered devices, such as household appliances and industrial equipment, more cost-effective and reliable.

AC waveforms have distinct properties that characterize their behavior:
1. Amplitude: The amplitude of an AC waveform represents the maximum value of the voltage or current. It indicates the strength of the electrical signal and is typically measured in volts (V) for voltage and amperes (A) for current.
2. Frequency: The frequency of an AC waveform is the number of complete cycles that occur within a unit of time, usually one second. It is measured in hertz (Hz) and is a crucial factor in determining the waveform's properties, such as the time required for one complete cycle and the energy content of the signal.

3. Phase: The phase of an AC waveform describes its position in time relative to another waveform of the same frequency. Phase differences between AC waveforms are important in applications such as power factor correction, synchronization of motors, and signal processing. Phase is typically measured in degrees (°) or radians.

In conclusion, AC offers several advantages over DC, including ease of transformation, generation, and distribution, as well as improved motor performance. Understanding the properties of AC waveforms, such as amplitude, frequency, and phase, is crucial for mastering electrical theory and excelling as a master electrician.

**Capacitors and inductors** are two essential components in AC circuits, each with unique characteristics and functions that influence voltage and current. Understanding their roles in various circuit configurations is vital for mastering electrical theory and becoming a skilled electrician.

1. Capacitors: Capacitors are passive components that store energy in an electric field when a voltage is applied across their terminals. They consist of two conductive plates separated by an insulating material called a dielectric. The primary characteristics of capacitors in AC circuits are:

- Impedance: Capacitive reactance ($X_c$) is the opposition offered by a capacitor to the flow of AC current. It is inversely proportional to the frequency (f) and capacitance (C) and is calculated as $X_c = 1/(2\pi f C)$.
- Phase: In a purely capacitive AC circuit, the voltage lags the current by 90 degrees. This means that the current reaches its peak value before the voltage does.
- Function: Capacitors are commonly used for filtering, energy storage, signal coupling, and tuning in AC circuits.

2. Inductors: Inductors are passive components that store energy in a magnetic field when a current flows through them. They consist of a coil of wire, usually wound around a magnetic core. The primary characteristics of inductors in AC circuits are:

- Impedance: Inductive reactance ($X_l$) is the opposition offered by an inductor to the flow of AC current. It is directly proportional to the frequency (f) and inductance (L) and is calculated as $X_l = 2\pi f L$.
- Phase: In a purely inductive AC circuit, the voltage leads the current by 90 degrees. This means that the voltage reaches its peak value before the current does.
- Function: Inductors are commonly used for filtering, energy storage, signal coupling, and tuning in AC circuits.

When capacitors and inductors are combined in AC circuits, they can create various configurations that influence voltage and current in different ways:

1. Series RLC circuits: In this configuration, the resistive (R), capacitive ($X_c$), and inductive ($X_l$) components are connected in series. The total impedance (Z) of the circuit is calculated as $Z = \sqrt{R^2 + (X_l - X_c)^2}$. The phase difference between voltage and current depends on the balance between $X_l$ and $X_c$.
2. Parallel RLC circuits: In this configuration, the resistive (R), capacitive ($X_c$), and inductive ($X_l$) components are connected in parallel. The total impedance (Z) of the circuit is determined using the reciprocal formula: $1/Z = 1/R + 1/jX_l - 1/jX_c$. In this case, the phase relationship between voltage and current is influenced by the relative magnitudes of R, $X_l$, and $X_c$.

In conclusion, capacitors and inductors play crucial roles in AC circuits, with unique functions and characteristics that influence voltage and current. A thorough understanding of these components and their impact on various circuit configurations is essential for mastering electrical theory and excelling as a master electrician.

**Impedance** is a crucial concept in AC circuits, as it influences how voltage and current behave within the circuit. It represents the total opposition a circuit offers to the flow of alternating current and accounts for both resistive and reactive components. In this discussion, we'll dive into the concept of impedance and learn how to calculate it in series and parallel circuits containing resistors, capacitors, and inductors.

1. Impedance in AC circuits: Impedance (Z) is a complex quantity that accounts for resistance (R), capacitive reactance (Xc), and inductive reactance (Xl). It is represented as $Z = R + j(Xl - Xc)$, where j is the imaginary unit. The magnitude of impedance is denoted by $|Z|$, and its angle (phase angle) is given by $\theta = \arctan((Xl - Xc)/R)$.
2. Impedance in series circuits: In a series circuit containing resistors, capacitors, and inductors, the total impedance is the vector sum of individual impedances. The real part of the total impedance is the sum of resistances, and the imaginary part is the difference between the inductive and capacitive reactances. The total impedance in a series circuit is given by $Z = R + j(Xl - Xc)$ and its magnitude by $|Z| = \sqrt{R^2 + (Xl - Xc)^2}$.
3. Impedance in parallel circuits: In a parallel circuit containing resistors, capacitors, and inductors, the total impedance is calculated using the reciprocal formula. For the real part, the total conductance (G) is the sum of individual conductances (1/R). For the imaginary part, the total susceptance (B) is the difference between capacitive and inductive susceptances $(1/jXc - 1/jXl)$. The total impedance in a parallel circuit is determined as $Z = 1/(G + jB)$ and its magnitude by $|Z| = 1/\sqrt{G^2 + B^2}$.

Understanding impedance and its impact on AC circuits is essential for analyzing and designing electrical systems. By learning how to calculate impedance in series and parallel circuits containing resistors, capacitors, and inductors, you'll develop a strong foundation in electrical theory, enhancing your ability to excel as a master electrician.

**Power factor** is a vital concept in AC circuits, as it directly influences power consumption and the efficiency of electrical systems. In this discussion, we'll explore the principles of power factor, its importance in AC circuits, and its impact on power consumption and efficiency.

1. Definition of power factor: Power factor (PF) is the ratio of active (real) power (P) to apparent power (S) in an AC circuit. It is a dimensionless value that ranges between -1 and 1. Mathematically, PF = P/S. Power factor represents the cosine of the angle between the current and voltage waveforms in the circuit. In other words, $PF = \cos(\theta)$, where $\theta$ is the phase angle between the current and voltage.
2. Types of power factors: Power factors are classified into three types: lagging, leading, and unity. A lagging power factor (PF < 1) occurs when the current lags behind the voltage, typical in inductive loads like motors and transformers. A leading power factor (PF > -1) is observed when the current leads the voltage, which is common in capacitive loads. A unity power factor (PF = 1) indicates that the current and voltage are in phase, maximizing power efficiency.

3. Importance of power factor: The power factor is critical in AC circuits because it impacts the efficiency of power transmission and distribution. A low power factor implies that the electrical system is consuming more reactive power, which contributes to higher losses in the form of heat and decreased efficiency. Therefore, maintaining a high power factor close to unity is essential to minimize power losses and optimize energy consumption.
4. Power factor correction: To improve the power factor, power factor correction devices like capacitors and synchronous motors are used. By introducing capacitors in parallel with inductive loads, the reactive power caused by the inductance is partially compensated, resulting in a higher power factor. Synchronous motors can also be used to adjust the power factor by varying their excitation current.

Understanding the principles of power factor and its impact on power consumption and efficiency is essential for designing and operating electrical systems. By maintaining a high power factor, you'll contribute to a more efficient and sustainable use of electrical energy, which is crucial for the success of a master electrician.

**Transformers** are essential components in electrical systems, playing a crucial role in voltage regulation, isolation, and impedance matching. In this discussion, we will describe the operation and applications of transformers and examine their significance in these areas.
1. Operation of transformers: Transformers are electromagnetic devices that transfer electrical energy between two or more coils through mutual induction. They consist of a primary coil, a secondary coil, and a magnetic core. By varying the number of turns in the primary and secondary coils, transformers can step-up or step-down voltage levels according to the application requirements. The voltage transformation ratio is determined by the ratio of the number of turns in the primary and secondary coils ($N1/N2$).
2. Voltage regulation: One of the primary applications of transformers is to regulate voltage levels in electrical systems. Step-up transformers increase the voltage, making it suitable for long-distance transmission with reduced power losses. Conversely, step-down transformers decrease the voltage to safe levels for distribution and end-user consumption. This voltage regulation helps maintain the stability and efficiency of power systems.
3. Isolation: Transformers also provide electrical isolation between circuits, ensuring safety and reducing the risk of electrical hazards. Isolation transformers are designed with a 1:1 turns ratio, meaning they do not alter the voltage but separate the primary and secondary circuits. This separation is crucial for protecting sensitive equipment from voltage surges, noise, and ground loops.
4. Impedance matching: Transformers are employed to match impedance between different circuits or devices, maximizing power transfer and minimizing signal reflections. In telecommunications, audio systems, and other applications, impedance matching is vital for ensuring optimal signal quality and efficient power utilization.

In summary, transformers are critical components in electrical systems, with important roles in voltage regulation, isolation, and impedance matching. As a master electrician, understanding their operation and applications will help you design, install, and maintain efficient and reliable power systems.

**Three-phase electrical systems** are a widely used method of power generation, transmission, and distribution in industrial and commercial applications. They consist of three sinusoidal AC voltages, each with the same frequency and amplitude but offset by 120 degrees in phase from one another. Understanding the principles of three-phase systems and their advantages is crucial for a master electrician working in power distribution and industrial environments.

1. Generation and transmission: Three-phase systems are generated by rotating a set of three coils within a magnetic field, producing three separate AC voltages with a phase difference of 120 degrees. This configuration makes power generation and transmission more efficient since it allows for smaller conductors and reduced power losses compared to single-phase systems.
2. Balanced loads: In a balanced three-phase system, the three phases equally share the load. This balance results in a constant power transfer throughout the cycle, reducing voltage fluctuations and ensuring stable power delivery. It also allows for the cancellation of neutral currents, making it possible to eliminate the neutral conductor in some applications.
3. Higher power capacity: Three-phase systems can deliver more power with fewer conductors compared to single-phase systems. This higher power capacity makes three-phase systems particularly well-suited for industrial and commercial applications, where large electrical loads are common.
4. Motors and drives: Three-phase systems are advantageous for operating electric motors and drives. Three-phase induction motors are more efficient, have a higher power-to-weight ratio, and provide smoother torque compared to single-phase motors. These characteristics make three-phase motors the preferred choice for industrial applications, where reliable and efficient motor operation is crucial.
5. Simplified transformer design: In three-phase systems, transformers can be designed with a simpler construction than their single-phase counterparts. This simplified design leads to reduced size, weight, and cost, as well as increased reliability.

In conclusion, three-phase electrical systems offer several advantages in power distribution and industrial applications, including improved efficiency, balanced load sharing, higher power capacity, and better performance of motors and drives. A comprehensive understanding of three-phase systems is essential for master electricians working in industrial and commercial environments.

**Electrical diagrams** are essential tools for master electricians, as they provide a visual representation of electrical circuits and systems, aiding in the design, installation, maintenance, and troubleshooting processes. There are several types of electrical diagrams, each serving a specific purpose. Here, we will introduce schematic, wiring, and block diagrams and discuss their uses in representing electrical circuits and systems.

1. Schematic diagrams: Schematic diagrams, also known as circuit diagrams, are a symbolic representation of electrical circuits. They use standardized symbols to represent various electrical components like resistors, capacitors, switches, and more. Schematic diagrams depict the arrangement of these components and the connections between them, allowing electricians to understand the circuit's functional design and analyze its performance. These diagrams are particularly useful for troubleshooting and designing complex circuits.

2. Wiring diagrams: Wiring diagrams provide a detailed representation of the physical connections and layout of an electrical system or circuit. They show the actual wiring paths, including wire colors and sizes, between various components and devices like outlets, switches, and junction boxes. Wiring diagrams are crucial for installation, maintenance, and repair work, as they help electricians to plan the routing of wires and ensure the connections are made correctly and safely.
3. Block diagrams: Block diagrams are a simplified representation of electrical systems, breaking them down into functional blocks or stages. Each block represents a specific function or operation within the system, and the connections between blocks indicate the flow of signals or power. Block diagrams are valuable for understanding the overall operation of complex systems, facilitating system design, and identifying potential issues or areas for improvement.

In summary, electrical diagrams are indispensable tools for master electricians, offering visual representations of electrical circuits and systems that aid in design, installation, troubleshooting, and maintenance tasks. Schematic, wiring, and block diagrams each serve a specific purpose and help electricians understand the functional design, physical connections, and overall operation of electrical systems. Familiarity with these diagrams and their interpretation is essential for any master electrician.

Electrical safety is a paramount concern for master electricians. Understanding key safety concepts, such as grounding, short circuits, and circuit protection devices like fuses and circuit breakers, is vital for ensuring the safe design, installation, and maintenance of electrical systems. Let's explore these concepts in more detail.

1. Grounding: Grounding is the process of connecting an electrical system or device to the earth or another conducting body, which serves as a reference point for voltage levels. This connection provides a safe path for excess current to flow in the event of a fault, such as a short circuit or lightning strike, thereby reducing the risk of electric shock, fire, or equipment damage. Grounding also helps stabilize voltage levels and improve the overall performance and safety of electrical systems.
2. Short circuits: A short circuit occurs when an unintended low-resistance path is established between two points in an electrical circuit with different voltage levels, resulting in a sudden surge of current. This can be caused by various factors, such as damaged insulation, loose connections, or the accidental contact of conductive materials. Short circuits can lead to electric shock, fires, or damage to electrical components. To minimize the risk of short circuits, electricians should ensure proper insulation, secure connections, and the use of appropriate protective devices.
3. Circuit protection devices: Fuses and circuit breakers are two common types of circuit protection devices that help prevent damage to electrical systems and components due to excessive current or short circuits.

a. Fuses: A fuse is a sacrificial device that contains a thin wire or strip designed to melt and break the circuit when the current exceeds a specified limit. When a fuse blows, it must be replaced with a new one of the appropriate rating to restore power to the circuit.
b. Circuit breakers: Circuit breakers are reusable devices that automatically interrupt the flow of current when it exceeds a specified limit. They work by using a bimetallic strip or an electromagnet to detect overcurrent conditions, causing the breaker to trip and open the circuit.

Circuit breakers can be reset after the fault has been resolved, making them a more convenient and environmentally friendly option than fuses.

In conclusion, understanding and applying key electrical safety concepts such as grounding, short circuits, and circuit protection devices is essential for master electricians. These concepts help ensure the safe operation of electrical systems, prevent electric shocks, fires, and equipment damage, and promote the well-being of both professionals and end-users.

# **National Electrical Code (NEC)**

The National Electrical Code (NEC) is a comprehensive set of guidelines and standards for the safe installation and maintenance of electrical systems in the United States. Developed by the National Fire Protection Association (NFPA) and updated every three years, the NEC is widely adopted across the country to ensure the safety of electrical installations in residential, commercial, and industrial settings. In this chapter, we'll provide an overview of the NEC, focusing on its key principles, structure, and relevance to master electricians.

1. Purpose: The primary objective of the NEC is to safeguard people and property from the hazards arising from the use of electricity. It serves as a baseline for the design, installation, and inspection of electrical systems, helping to minimize the risks of electric shock, fire, and other electrical hazards.
2. Structure: The NEC is organized into several chapters, articles, and sections, addressing various aspects of electrical installations. Some of the main topics covered in the code include:
    - Wiring and protection: grounding, overcurrent protection, branch circuits, and feeders
    - Wiring methods and materials: raceways, cables, conductors, and boxes
    - Equipment for general use: receptacles, switches, lighting, appliances, and motors
    - Special occupancies: hazardous locations, healthcare facilities, and theaters
    - Special equipment: solar photovoltaic systems, fuel cell systems, and energy storage systems
    - Special conditions: emergency systems, standby power, and fire alarm systems
    - Communications systems: telephone, radio, and television installations
3. Code updates and local adoption: The NEC is updated every three years to incorporate new technologies, materials, and best practices. However, the adoption of the latest version of the NEC is determined by local jurisdictions, which may have their own amendments or additional requirements. As a master electrician, it's important to be familiar with the version of the NEC adopted in your area and stay up-to-date with any changes.
4. Compliance and enforcement: Compliance with the NEC is typically enforced through local building codes and electrical inspection authorities. Master electricians must adhere to the NEC when designing, installing, and maintaining electrical systems to ensure their work meets the required safety standards. Non-compliance with the NEC can result in fines, penalties, or even the denial of permits.

In summary, the National Electrical Code is a vital resource for master electricians, providing a comprehensive set of guidelines and standards for the safe installation and maintenance of electrical systems. Familiarity with the NEC and its local adaptations is essential for ensuring the safety and compliance of your electrical work.

The National Electrical Code (NEC), also known as NFPA 70, is a comprehensive set of guidelines designed to ensure the safe installation and operation of electrical systems in

residential, commercial, and industrial settings. The NEC is organized into a hierarchical structure that includes chapters, articles, sections, and subsections.
1. Chapters: The NEC is divided into nine chapters, each focusing on a specific aspect of electrical installation or equipment. These chapters cover a wide range of topics, from wiring and protection to specialized equipment and occupancies.
2. Articles: Each chapter consists of multiple articles that address specific subjects within the broader context of the chapter. For example, within the "Wiring and Protection" chapter, you'll find articles dedicated to grounding, branch circuits, feeders, and overcurrent protection, among other topics.
3. Sections: Articles are further divided into sections, which provide detailed guidelines and requirements for the subject matter of the article. Sections usually contain an alphanumeric code, such as 210.52, to facilitate easy reference. These codes represent the chapter, article, and section number, respectively.
4. Subsections: Sections may be subdivided into subsections, which clarify or expand upon the requirements outlined in the section. Subsections are typically denoted with a letter or a series of letters, such as (A), (B), or (C). These subsections often contain additional information, exceptions, or notes that provide further guidance for specific situations.

The NEC's organization and structure make it easier for users to locate and reference the information they need. It is important to become familiar with this layout to effectively navigate the code and ensure compliance with its requirements. As a master electrician, understanding the organization of the NEC is essential for successful completion of the Master Electrician exam and for performing safe, code-compliant electrical work in the field.

To better understand and apply the NEC, it's crucial to become familiar with key terms and definitions that are frequently used throughout the code. Here are explanations for some essential NEC terms:
1. Grounded Conductor: A grounded conductor is a system or circuit conductor that is intentionally connected to the earth, typically through a grounding electrode. This connection helps stabilize voltage levels and provides a path for fault current to flow, ensuring the safety of electrical systems.
2. Grounding Electrode: A grounding electrode is a conductive object that establishes a direct electrical connection to the earth. It serves to dissipate fault currents, lightning strikes, and static charges into the ground, helping to prevent electrical shock and equipment damage. Common types of grounding electrodes include metal water pipes, ground rods, and metallic building frames.
3. Branch Circuit: A branch circuit is a portion of an electrical system that extends from the final overcurrent protection device, such as a circuit breaker or fuse, to the outlets or devices it supplies. Branch circuits can be dedicated to a single device or supply multiple devices, depending on the application. They are designed to carry the current required for the connected load and must meet specific NEC requirements for wiring size, overcurrent protection, and other factors.
4. Feeder: A feeder is an electrical circuit that extends from the service equipment, a branch circuit overcurrent protection device, or another feeder overcurrent protection device to the distribution equipment or final branch circuits. Feeders are used to transmit electrical power between different parts of a building or facility and are typically larger in size than

branch circuits. They must meet NEC requirements for wire sizing, overcurrent protection, and other factors, depending on the intended application.

By mastering these and other key terms, you'll be better equipped to navigate and apply the NEC in your work as a master electrician. Understanding these terms is essential for passing the Master Electrician exam and ensuring safe, code-compliant installations.

The NEC dedicates a significant portion of its content to wiring and protection, ensuring the safety and proper functioning of electrical systems. Some of the essential requirements for grounding, bonding, overcurrent protection, and conductor sizing are discussed below:

1. Grounding: Grounding is the intentional connection of an electrical system or circuit to the earth. This connection helps stabilize voltage levels, provides a path for fault currents, and prevents electrical shock. The NEC mandates specific grounding requirements, such as grounding electrode conductors' sizing and types of grounding electrodes that can be used. Refer to NEC Article 250 for detailed information on grounding requirements.
2. Bonding: Bonding is the process of connecting metallic, non-current-carrying parts of electrical equipment to ensure electrical continuity and a low-impedance path for fault currents. Proper bonding prevents potential differences that could lead to electrical shock hazards. NEC Article 250 provides guidelines on bonding methods and requirements, such as bonding jumpers and the use of bonding bushings.
3. Overcurrent Protection: Overcurrent protection devices, like fuses and circuit breakers, are crucial for safeguarding electrical systems from excessive current that could result in equipment damage or fire. The NEC outlines specific requirements for selecting and installing overcurrent protection devices. For instance, overcurrent protection must be sized according to the ampacity of the conductors they protect, as specified in NEC Article 240.
4. Conductor Sizing: Choosing the right conductor size is vital to ensure the safe and efficient operation of electrical circuits. The NEC provides guidelines for selecting proper conductor sizes based on factors such as continuous and non-continuous loads, ambient temperature, and conductor insulation type. NEC Article 310 and its accompanying tables are essential resources for determining conductor sizes and ampacities.

By adhering to the NEC's requirements for wiring and protection, you can ensure that electrical installations are safe, efficient, and code-compliant. Familiarizing yourself with these concepts is crucial for passing the Master Electrician exam and succeeding in your career as a master electrician.

The NEC dedicates a significant portion of its content to wiring and protection, ensuring the safety and proper functioning of electrical systems. Some of the essential requirements for grounding, bonding, overcurrent protection, and conductor sizing are discussed below:

1. Grounding: Grounding is the intentional connection of an electrical system or circuit to the earth. This connection helps stabilize voltage levels, provides a path for fault currents, and prevents electrical shock. The NEC mandates specific grounding requirements, such as grounding electrode conductors' sizing and types of grounding electrodes that can be used. Refer to NEC Article 250 for detailed information on grounding requirements.
2. Bonding: Bonding is the process of connecting metallic, non-current-carrying parts of electrical equipment to ensure electrical continuity and a low-impedance path for fault currents. Proper bonding prevents potential differences that could lead to electrical shock

hazards. NEC Article 250 provides guidelines on bonding methods and requirements, such as bonding jumpers and the use of bonding bushings.
3. Overcurrent Protection: Overcurrent protection devices, like fuses and circuit breakers, are crucial for safeguarding electrical systems from excessive current that could result in equipment damage or fire. The NEC outlines specific requirements for selecting and installing overcurrent protection devices. For instance, overcurrent protection must be sized according to the ampacity of the conductors they protect, as specified in NEC Article 240.
4. Conductor Sizing: Choosing the right conductor size is vital to ensure the safe and efficient operation of electrical circuits. The NEC provides guidelines for selecting proper conductor sizes based on factors such as continuous and non-continuous loads, ambient temperature, and conductor insulation type. NEC Article 310 and its accompanying tables are essential resources for determining conductor sizes and ampacities.

By adhering to the NEC's requirements for wiring and protection, you can ensure that electrical installations are safe, efficient, and code-compliant. Familiarizing yourself with these concepts is crucial for passing the Master Electrician exam and succeeding in your career as a master electrician.

Wiring methods and materials: Cover the NEC guidelines for selecting and installing raceways, cables, conductors, junction boxes, and other wiring components.

The National Electrical Code (NEC) provides comprehensive guidelines for selecting and installing various wiring components to ensure safe and reliable electrical systems. The key aspects of wiring methods and materials are outlined below:

1. Raceways: Raceways are enclosed channels designed to protect and route electrical conductors. The NEC specifies different types of raceways, such as conduit, tubing, and cable trays, along with their appropriate applications and installation methods. Refer to NEC Article 300 for general wiring requirements and Articles 342-392 for specific raceway types and their provisions.
2. Cables: Cables are pre-assembled groups of conductors that are protected by a non-metallic or metallic sheath. The NEC provides guidelines for choosing the right cable type, such as non-metallic sheathed (NM) cable, armored cable (AC), or metal-clad (MC)

cable, depending on factors like the environment, load type, and voltage rating. NEC Articles 320-340 offer detailed information on cable types, installation methods, and restrictions.
3. Conductors: The NEC outlines requirements for selecting the right conductor material, size, insulation type, and color-coding. It is crucial to choose conductors that can safely handle the current and withstand environmental conditions. NEC Article 310 provides guidance on conductor sizing, ampacity, and insulation requirements, while Article 200 covers identification and color-coding rules.
4. Junction Boxes: Junction boxes house splices, taps, or terminations of conductors and provide access for maintenance. The NEC mandates specific requirements for junction box sizing, material, and installation, ensuring that the box can accommodate the conductors and devices it houses without causing overheating or damage. Refer to NEC Article 314 for more information on junction box requirements.
5. Other Wiring Components: The NEC also provides guidelines for other wiring components, such as switches, receptacles, and connectors, to ensure they are appropriate for the intended application and properly installed. For instance, NEC Article 404 covers switch installation and usage, while Article 406 discusses receptacle requirements.

By following the NEC's guidelines for wiring methods and materials, you can create safe, efficient, and code-compliant electrical installations. Familiarity with these principles is essential for passing the Master Electrician exam and excelling in your electrical career.

The National Electrical Code (NEC) establishes standards for the safe installation and use of various electrical equipment, including receptacles, switches, lighting fixtures, appliances, and motors. Understanding these provisions is crucial for electricians to ensure safety and compliance.
1. Receptacles: NEC Article 406 specifies requirements for receptacles, such as proper grounding, tamper-resistant designs, and ground-fault circuit interrupter (GFCI) protection in wet or damp locations. Additionally, the NEC mandates proper spacing and installation height of receptacles to ensure convenient and safe access.
2. Switches: Article 404 of the NEC outlines provisions for switches, including proper rating, grounding, and installation of switch boxes. For instance, the NEC requires that switches controlling lighting loads be rated for the specific type of light fixture and installed in an accessible location.
3. Lighting Fixtures: The NEC regulates the installation and use of lighting fixtures in Article 410. Requirements include proper support, clearances from combustible materials, and guidelines for specific environments like wet locations. The NEC also addresses energy conservation and safety aspects, such as lamp wattage limitations and the use of energy-efficient lighting.
4. Appliances: NEC Article 422 covers the installation and use of appliances, such as water heaters, air conditioners, and cooking equipment. The code specifies requirements for disconnecting means, overcurrent protection, and proper grounding. For appliances with specific safety concerns, like electric space heaters or swimming pool equipment, the NEC offers additional guidelines to ensure safe operation.
5. Motors: Article 430 of the NEC deals with motor installation, protection, and control. It outlines requirements for motor sizing, overcurrent protection, grounding, and short-

circuit protection. The NEC also addresses the safe installation and use of motor controllers and disconnecting means.

By adhering to the NEC provisions for equipment installation and usage, you can ensure safe and reliable operation while avoiding potential hazards. Familiarity with these requirements is essential for success in the Master Electrician exam and your professional career in the electrical industry.

The NEC includes specific provisions for special occupancies, equipment, and conditions to address unique safety concerns and requirements. Some of these include hazardous locations, healthcare facilities, solar photovoltaic systems, and emergency systems.

1. Hazardous Locations: NEC Articles 500 to 517 cover the requirements for hazardous locations, such as areas with flammable gases, combustible dust, or ignitable fibers. The code classifies these locations based on the type of hazard, likelihood of occurrence, and material properties. It also specifies the types of electrical equipment and wiring methods allowed in such areas to minimize the risk of ignition.
2. Healthcare Facilities: NEC Article 517 focuses on the safety requirements for healthcare facilities, including hospitals, nursing homes, and outpatient clinics. The code addresses topics such as isolated power systems, ground-fault protection, and proper wiring methods for patient care areas. It also requires the installation of essential electrical systems to ensure the continuous operation of life support and critical care equipment during power outages.
3. Solar Photovoltaic Systems: Article 690 of the NEC provides guidelines for the installation and safety of solar photovoltaic (PV) systems. These provisions include requirements for system grounding, overcurrent protection, disconnecting means, and wiring methods. The NEC also addresses issues related to the proper labeling and interconnection of solar PV systems with the electrical grid.
4. Emergency Systems: NEC Article 700 establishes requirements for emergency systems designed to provide power during utility outages. These provisions cover topics like system design, installation, maintenance, and testing. The NEC also specifies requirements for emergency lighting, exit signs, and other life safety equipment.

Other special occupancies, equipment, and conditions addressed in the NEC include:
- Electric vehicle charging systems (Article 625)
- Swimming pools and fountains (Article 680)
- Temporary installations (Article 590)
- Agricultural buildings (Article 547)
- Marina and boatyard installations (Article 555)

Understanding the specific requirements for these unique situations is crucial for electricians to ensure safety and compliance in various electrical installations. Familiarizing yourself with these provisions will also help you succeed in the Master Electrician exam and enhance your expertise in the electrical field.

The NEC addresses the installation and safety requirements for communication systems, including telephone, radio, television, and other communication installations, in Articles 800 to 830. These provisions aim to ensure the safe and reliable operation of such systems while minimizing interference and hazards.

Telephone and Data Communications: Article 800 covers the rules for the installation and protection of telephone, data, and broadband communication circuits. These guidelines include proper grounding and bonding, cable routing, separation from power conductors, and fire protection measures. The article also addresses the use of surge protectors and cable types for different environments.

Radio and Television Installations: Article 810 provides guidance for the safe installation and grounding of radio and television receiving and transmitting equipment, such as antennas, satellite dishes, and supporting structures. The NEC emphasizes the importance of proper grounding to minimize the risk of damage from lightning and to ensure the safety of personnel.

Network-Powered Broadband Communications Systems: Article 830 pertains to the installation of network-powered broadband communication systems, which may include internet, telephone, or video services. The provisions in this article cover topics such as cable types, grounding and bonding, and separation from power conductors. The NEC also emphasizes the need for proper grounding to protect against voltage surges and lightning.

Optical Fiber Cables and Raceways: Article 770 deals with the installation of optical fiber cables and raceways used in communication systems. The guidelines in this article focus on the proper selection and installation of cables, cable routing, and fire protection measures. Additionally, the NEC addresses the use of optical fiber cables in hazardous locations.

By understanding the NEC rules and guidelines for various communication systems, you will be better equipped to ensure safe and reliable installations. Familiarity with these provisions is also essential for passing the Master Electrician exam and advancing your expertise in the electrical field.

The National Electrical Code (NEC) is a living document that undergoes regular updates to stay current with evolving technology, new materials, and best practices in the electrical industry. The NEC's update process follows a three-year revision cycle, which aims to ensure that the code remains relevant and effective in promoting electrical safety.

1. Three-Year Revision Cycle: The NEC is updated every three years by the National Fire Protection Association (NFPA), the organization responsible for the code. The revision process begins with the submission of proposals for changes, which can be made by anyone with an interest in electrical safety, such as engineers, electricians, manufacturers, and regulatory authorities. The proposed changes are then reviewed by various committees composed of industry experts.
2. Public Input and Comment: During the revision process, public input is encouraged, and stakeholders can submit their comments on the proposed changes. The committees consider this feedback and may make adjustments to the proposals based on the input received. Once the committees have reviewed all the proposals and comments, they vote on the final set of changes to be incorporated into the updated NEC.
3. Publication and Adoption: After the revision process is complete, the updated NEC is published and becomes available for adoption by local jurisdictions. It is essential to note

that the NEC is not a law in itself; it serves as a model code that must be adopted by local or state governments to become legally enforceable.
4. Local Adoption and Enforcement: The adoption of the NEC varies among states, counties, and cities. Some jurisdictions adopt the code as is, while others may make modifications or amendments to better suit their local conditions and requirements. Once adopted, the NEC is enforced by local building and electrical inspectors, who review electrical installations for compliance with the code.

By understanding the NEC update process and how it is adopted and enforced at the local level, electricians and other professionals can stay informed about the latest safety requirements and best practices in the electrical industry.

Navigating and interpreting the NEC can be challenging, especially for those new to the code. However, with practice and the right strategies, you can effectively use the NEC as a reference and apply its requirements in real-world scenarios. Here are some tips for understanding and utilizing the NEC:

1. Familiarize yourself with the structure: The NEC is organized into chapters, articles, sections, and subsections. Begin by understanding this structure and how the content is organized, so you can easily locate the relevant information when needed.
2. Use the index and table of contents: Make use of the index and table of contents provided in the NEC. These tools can help you quickly find the specific sections or articles you are looking for, saving time and frustration.
3. Understand the language: The NEC uses specific terminology and language that may be unfamiliar at first. Take the time to learn the meanings of essential terms and phrases, as this will make it easier to understand the code's requirements.
4. Read the informational notes: Throughout the NEC, you will find informational notes that provide clarification, background, or additional information about a specific requirement. Read these notes, as they can be valuable in helping you understand the context and rationale behind the code's provisions.
5. Cross-reference related sections: When studying a particular requirement, be sure to check for any related sections or articles that may provide additional information or clarification. This will help you gain a more comprehensive understanding of the topic at hand.
6. Practice with real-world scenarios: Apply the NEC's requirements to real-world examples and case studies to reinforce your understanding and develop your ability to interpret and apply the code in practical situations.
7. Seek help from experienced professionals: If you are unsure about a specific requirement or have difficulty interpreting the code, do not hesitate to consult with experienced professionals or colleagues who may have more expertise in the area.
8. Stay up-to-date with changes: Remember that the NEC is updated every three years. Stay informed about the latest revisions and their implications to ensure your knowledge remains current and accurate.

By following these tips and strategies, you can effectively navigate and interpret the NEC, allowing you to apply its requirements confidently and accurately in real-world situations.

Master electricians should be aware of common NEC violations and pitfalls to ensure the safety and compliance of their work. Here are some frequent mistakes and misunderstandings that should be avoided:
1. Insufficient conductor sizing: Selecting a conductor size that is too small for the load can lead to overheating and fire hazards. Always use the appropriate conductor size based on the load calculations and NEC requirements.
2. Overcrowded junction boxes: Overcrowding junction boxes can cause heat buildup and increase the risk of electrical fires. Ensure that your junction boxes are sized correctly and comply with NEC's box fill requirements.
3. Inadequate grounding and bonding: Grounding and bonding are essential for electrical safety. Make sure you follow the NEC's guidelines on grounding electrode systems, bonding of metal parts, and proper connections to avoid potential hazards.
4. Misuse of extension cords: Extension cords should only be used for temporary purposes and not as a substitute for permanent wiring. Using extension cords improperly can lead to electrical fires and other safety hazards.
5. Incorrect circuit breaker sizing: Circuit breakers must be correctly sized to protect conductors from overcurrents. Using a circuit breaker that is too large for the conductor can result in overheating and potential fire hazards.
6. Improper GFCI and AFCI protection: The NEC requires Ground Fault Circuit Interrupter (GFCI) and Arc Fault Circuit Interrupter (AFCI) protection in specific locations to prevent electrical shocks and fires. Ensure that you follow these requirements and install the appropriate devices as needed.
7. Inappropriate use of wiring methods and materials: Always choose the correct wiring methods and materials for the specific application, and make sure they comply with the NEC's guidelines for the environment and conditions in which they will be used.
8. Ignoring local amendments: Keep in mind that local jurisdictions may have their own amendments to the NEC. Be aware of these changes and ensure your work complies with both the NEC and any local requirements.
9. Failing to follow NEC updates: The NEC is updated every three years, and staying informed about the latest changes is crucial. Make sure you stay up-to-date with the most recent revisions to avoid violations and maintain the highest level of safety.

By being aware of these common NEC violations and pitfalls, master electricians can avoid costly mistakes, ensure compliance, and provide safe and reliable electrical installations.

Electrical inspectors play a crucial role in maintaining the safety and compliance of electrical installations by ensuring adherence to the NEC and local code requirements. They act as a vital link between electricians, contractors, and the local jurisdiction.
The primary responsibilities of electrical inspectors include:
1. Reviewing plans and permits: Inspectors review the plans submitted by electricians and contractors to ensure that the proposed work meets the NEC and local code requirements. They verify that the proper permits have been obtained before any work begins.
2. Inspecting electrical work: Inspectors conduct site visits to examine the ongoing electrical work at various stages of completion. They ensure that the installation follows the approved plans and complies with the NEC and any local amendments.

3. Identifying code violations: If an inspector finds any code violations, they document the issues and inform the electrician or contractor. The responsible party must then correct the violations and have the work re-inspected before it can be approved.
4. Approving the installation: Once the electrical work has been completed and complies with all code requirements, the inspector issues a final approval, allowing the project to move forward to the next phase or completion.

The permitting process is essential for ensuring that electrical work is performed safely and in compliance with the NEC and local codes. It involves obtaining the necessary permits from the local jurisdiction before beginning any electrical work. The process typically includes:

1. Submitting an application: The electrician or contractor submits a permit application, including detailed plans of the proposed electrical work, to the local permitting office.
2. Plan review: The permitting office reviews the submitted plans and checks them for compliance with the NEC and local code requirements.
3. Issuing the permit: If the plans are approved, the permitting office issues the necessary permits, allowing the work to proceed.
4. Inspections: As the work progresses, the electrical inspector conducts periodic site visits to ensure compliance with the approved plans and code requirements.
5. Final approval: After the work is completed and has passed all required inspections, the inspector issues a final approval or certificate of occupancy, signifying that the installation meets all code requirements.

Understanding the role of electrical inspectors and the permitting process is essential for ensuring that electrical work is completed safely, in compliance with the NEC, and with the proper permits in place.

# Safety

The Safety Chapter serves as a comprehensive guide to understanding the fundamental principles and practices of electrical safety. This chapter emphasizes the importance of adhering to the National Electrical Code (NEC) requirements, as well as local codes, to ensure the safe installation and operation of electrical systems.

In this chapter, we will explore key electrical safety concepts, such as grounding, short circuits, and circuit protection devices like fuses and circuit breakers. We will also discuss the proper use of personal protective equipment (PPE) and safe work practices to minimize the risk of electrical hazards, such as electric shock, arc flash, and fire.

Furthermore, the chapter will delve into the role of electrical inspectors and the permitting process, highlighting their significance in maintaining safe and compliant electrical installations. We will also address common NEC violations and pitfalls, helping you to avoid frequent mistakes and misunderstandings in electrical work.

By the end of this chapter, you will have a solid understanding of the essential aspects of electrical safety, enabling you to confidently implement safe practices in your electrical projects and ensure the well-being of both people and property.

**Electrical safety fundamentals** are essential for anyone working with or around electrical systems. Understanding the basic principles of electrical safety can help prevent accidents, injuries, and fatalities. The following key concepts are crucial for ensuring a safe working environment when dealing with electricity:

1. Electricity and its risks: Electricity is the flow of electrons through a conductive material. When not handled properly, it can pose significant risks, such as electric shock, electrocution, burns, arc flash, and fires. Understanding these hazards is the first step toward preventing them.
2. Insulation and isolation: Insulation is the use of non-conductive materials to prevent the flow of electricity. Isolation involves separating electrical circuits or equipment from other conductive materials to prevent unintentional contact with electricity. Both insulation and isolation are vital for protecting individuals from electrical hazards.
3. Voltage, current, and resistance: Voltage is the force that pushes electrons through a conductor, while current is the flow of electrons. Resistance is a material's opposition to the flow of electric current. It's essential to understand these concepts as they help determine the potential hazards of electrical systems and the appropriate safety measures to be taken.
4. Path to ground: Electricity always seeks the path of least resistance to ground. The human body can become part of this path if it comes into contact with live electrical components. Ensuring that electrical systems are grounded and that there's no potential for human contact with live parts is critical for safety.
5. Safety guidelines and regulations: Adhering to established safety guidelines, such as the National Electrical Code (NEC), OSHA standards, and local regulations, is vital for

maintaining a safe work environment. These guidelines are in place to minimize the risks associated with electricity and ensure that all electrical installations and maintenance work are conducted safely.
6. Safe work practices: Proper training, using the right tools and equipment, and following established procedures can significantly reduce the risks associated with electrical work. Always practice good safety habits, such as de-energizing circuits before working on them, using lockout/tagout procedures, and wearing appropriate personal protective equipment (PPE).

Understanding and applying these electrical safety fundamentals are critical for protecting yourself and others from the dangers associated with electricity. By being aware of the risks and following established safety guidelines, you can help ensure a safe working environment for everyone involved.

**Grounding and bonding** are fundamental concepts in electrical safety, playing critical roles in preventing electrical hazards and ensuring the proper functioning of electrical systems. Let's delve into these concepts and their associated NEC requirements.
1. Grounding: Grounding is the process of connecting an electrical system or equipment to the earth, providing a pathway for electrical current to dissipate safely in the event of a fault. Grounding helps stabilize voltage levels, prevents the buildup of static electricity, and reduces the risk of electric shock and equipment damage. The NEC outlines various grounding requirements, such as proper grounding electrode systems, grounding conductors, and grounding connections.
2. Bonding: Bonding refers to the intentional connection of conductive materials, such as metallic equipment enclosures, to create a low-impedance path for fault current to flow safely to the ground. Bonding ensures that all metallic components in an electrical system are at the same electrical potential, reducing the risk of electric shock and facilitating the operation of overcurrent protection devices in the event of a fault. The NEC contains requirements for bonding, including proper sizing and installation of bonding conductors, as well as guidelines for bonding various components, such as raceways, enclosures, and equipment.
3. Grounding and bonding relationship: Both grounding and bonding are essential for electrical safety, and their roles are complementary. Grounding helps prevent electric shock by providing a low-resistance path for fault current to dissipate, while bonding ensures that all conductive materials are at the same electrical potential, minimizing the risk of current flowing through unintended paths, such as the human body.
4. NEC requirements: The NEC provides detailed requirements for grounding and bonding in various articles, such as Article 250 (Grounding and Bonding), which outlines general requirements, and other articles specific to different types of electrical installations. It's crucial to follow these NEC guidelines to ensure the safe operation of electrical systems and prevent electrical hazards.

By understanding the concepts of grounding and bonding, their roles in electrical safety, and the associated NEC requirements, you'll be better equipped to ensure the safe design, installation, and maintenance of electrical systems. Remember to always consult the latest edition of the NEC and follow local codes and regulations for specific requirements in your jurisdiction.

Circuit protection devices are essential components in electrical systems, designed to safeguard against electrical hazards such as overcurrent, short circuits, and ground faults. Fuses, circuit breakers, and ground fault circuit interrupters (GFCIs) are common types of circuit protection devices that serve distinct purposes in maintaining electrical safety. Let's explore each of these devices and their roles in protecting against electrical hazards.

1. Fuses: A fuse is a sacrificial device that contains a small, conductive element that melts when the current flowing through it exceeds a predetermined threshold. By melting, the fuse breaks the circuit, thereby preventing the flow of excessive current that could cause damage to equipment or pose a fire risk. Fuses are simple, reliable, and inexpensive, but they must be replaced after they have interrupted an overcurrent condition.
2. Circuit breakers: A circuit breaker is a reusable, electromechanical device that automatically interrupts the flow of current in a circuit when it detects an overcurrent condition. Circuit breakers can be manually reset after they have tripped, allowing them to be used multiple times. They come in various types, such as thermal, magnetic, or a combination of both, designed to respond to different types of overcurrent situations.
3. Ground fault circuit interrupters (GFCIs): GFCIs are specialized devices designed to protect people from electric shock by detecting small imbalances in the flow of current between the hot and neutral conductors. When a ground fault is detected, the GFCI quickly interrupts the circuit, reducing the risk of electric shock. GFCIs are typically installed in areas where water or moisture is present, such as bathrooms, kitchens, and outdoor outlets, as these locations pose a higher risk of electric shock.

Each of these circuit protection devices contributes to electrical safety by performing specific functions to mitigate electrical hazards. Fuses and circuit breakers protect against overcurrent and short circuits, while GFCIs guard against ground faults that can cause electric shock. By understanding the purpose and function of these devices, you can better appreciate their importance in maintaining safe electrical systems. Always adhere to the guidelines outlined in the NEC and local codes when installing and maintaining these devices to ensure the highest level of safety.

Adopting safe work practices when working with electricity is crucial in minimizing risks and ensuring the well-being of everyone involved. Implementing safety protocols, proper procedures, and utilizing appropriate tools and equipment can help prevent accidents and injuries. Here are some essential safety practices to follow when working with electricity:

1. Lockout/Tagout (LOTO) procedures: LOTO is a safety procedure that helps ensure that electrical equipment is properly de-energized and isolated from its energy source before maintenance or repair work begins. This procedure involves placing locks and tags on the energy-isolating devices, such as circuit breakers or disconnect switches, to prevent unauthorized re-energization. Workers should be trained in LOTO procedures and follow them diligently to prevent accidents.
2. Maintaining a safe distance: When working with electricity, it's essential to maintain a safe distance from energized parts and conductors. This distance, also known as the "minimum approach distance," depends on the voltage level of the equipment being worked on. Workers should be familiar with these distances and adhere to them to prevent electrical shock, arc flash, or other hazards.

3. Using appropriate tools and equipment: Utilizing insulated tools and equipment specifically designed for electrical work is vital in ensuring safety. These tools help to reduce the risk of accidental contact with energized components. Personal protective equipment (PPE) such as insulated gloves, safety glasses, and arc-rated clothing should be worn when working with or near electricity to minimize the risk of injury.
4. Verifying de-energized conditions: Before working on electrical equipment, always verify that it is de-energized using a suitable voltage tester or multimeter. Never assume that equipment is de-energized just because it has been locked out or tagged. This extra step can help prevent electrical shock or other accidents caused by unexpected energization.
5. Understanding and following safety guidelines: Familiarize yourself with safety guidelines and best practices, such as those provided by the National Electrical Code (NEC), the Occupational Safety and Health Administration (OSHA), and other local regulations. Following these guidelines can help ensure that you are taking the necessary precautions and adopting the proper procedures to maintain a safe working environment.

By incorporating these safe work practices and diligently following established procedures, you can create a safer environment for yourself and others when working with electricity. Remember, safety should always be the top priority in any electrical work, and it is the responsibility of each worker to contribute to a safe working atmosphere.

Understanding and addressing common electrical hazards is crucial in maintaining a safe work environment. Some of the typical electrical hazards include electric shock, arc flash, and fires. Let's discuss these hazards and explore strategies to identify and prevent them.

1. Electric shock: Electric shock occurs when a person comes into contact with an energized electrical component, causing electric current to flow through their body. This can result in injury or even death. To identify and prevent electric shock:
- Inspect wiring, equipment, and tools for damage before use.
- Use insulated tools and wear appropriate personal protective equipment (PPE).
- Maintain a safe distance from energized parts and follow minimum approach distances.
- Implement lockout/tagout (LOTO) procedures when working on electrical equipment.
- Never bypass safety devices, such as ground fault circuit interrupters (GFCIs).
2. Arc flash: An arc flash is a sudden release of electrical energy caused by an electrical fault, such as a short circuit or an equipment failure. It can result in severe burns, hearing damage, and other injuries. To identify and prevent arc flash hazards:
- Conduct an arc flash risk assessment to determine the potential hazard level and select appropriate PPE.
- Use arc-resistant equipment and enclosures where possible.
- Establish an electrically safe working condition by de-energizing equipment before working on it.
- Adhere to proper lockout/tagout procedures and follow safety guidelines for working near energized parts.
- Implement regular maintenance and inspection programs to identify potential equipment issues before they lead to an arc flash.
3. Electrical fires: Electrical fires can result from overloaded circuits, damaged wiring, or faulty equipment. To identify and prevent electrical fires:

- Follow the National Electrical Code (NEC) requirements for proper installation and maintenance of electrical systems.
- Use circuit protection devices, such as fuses and circuit breakers, to prevent overloading and overheating.
- Regularly inspect wiring, outlets, and equipment for signs of wear, damage, or loose connections.
- Avoid using damaged or frayed cords and never overload outlets or extension cords.
- Implement a preventative maintenance program to identify and address potential issues before they lead to a fire.

By being aware of these common electrical hazards and implementing the strategies mentioned above, you can significantly reduce the risk of accidents and injuries while working with electricity. Remember, safety should always be the top priority when working with electrical systems, and it is essential to stay informed and vigilant in recognizing and addressing potential hazards.

Role of electrical inspectors and permitting process: Electrical inspectors play a vital role in ensuring compliance with safety regulations and the NEC. They are responsible for examining electrical installations in residential, commercial, and industrial settings to confirm that they meet the necessary safety standards. The permitting process is a crucial aspect of this role, as it helps maintain accountability and traceability. Obtaining permits for electrical work is essential because:

- It verifies that the work complies with the NEC and other applicable codes.
- It ensures that electrical installations are safe and reduce the risk of hazards, such as fires and electrocution.
- It provides a record of the work performed, which can be beneficial for future modifications, maintenance, or inspections.
- It may be required by law, and failure to obtain a permit can result in fines or other legal consequences.

Common NEC violations and safety pitfalls: Master electricians should be aware of frequent mistakes and misunderstandings related to the NEC and electrical safety to avoid them. Some common issues include:

1. Improper grounding and bonding: Ensure proper grounding and bonding connections to create an effective fault current path, reducing the risk of shock and equipment damage.
2. Inadequate circuit protection: Install the correct size of fuses and circuit breakers to prevent overloading and overheating, which can lead to fires and equipment damage.
3. Insufficient workspace clearance: Maintain adequate clearance around electrical equipment, as specified by the NEC, to provide safe access for inspection, maintenance, and repair.
4. Incorrect wire sizing: Use the appropriate wire size for the circuit load to prevent overheating and potential fires.
5. Overcrowded junction boxes: Follow the NEC's guidelines on box fill calculations to avoid overcrowding, which can cause overheating, short circuits, and other hazards.
6. Misuse of extension cords: Use extension cords only for temporary purposes and not as a permanent wiring solution.

7. Non-compliant outdoor installations: Ensure that outdoor electrical installations are compliant with the NEC, including using weatherproof enclosures and GFCI protection where required.

By staying informed about common NEC violations and safety pitfalls and adhering to the guidelines and requirements, master electricians can significantly reduce the risk of accidents, injuries, and property damage. Always prioritize safety and follow best practices when working with electrical systems.

Training and qualifications: Proper training and qualifications are crucial for electrical workers to ensure safety and compliance with the NEC and other regulations. Obtaining the necessary certifications and participating in continuing education programs help electrical workers stay current with industry standards, new technologies, and safety practices. Some key points include:

- Apprenticeships: Hands-on experience and training under the supervision of a licensed electrician provide foundational knowledge and skills for aspiring electricians.
- Licensing and certification: Obtaining a license demonstrates a professional's competency in electrical work, compliance with safety regulations, and adherence to industry standards.
- Continuing education: Regular participation in training courses and workshops helps electrical professionals stay up-to-date with the latest safety practices, code changes, and advancements in the field.

Electrical safety in special environments: Different environments may require unique safety considerations and practices. Here are some examples of special environments and their associated safety measures:

1. Hazardous locations: In areas where flammable or combustible materials are present, such as chemical plants or refineries, electrical installations must adhere to specific safety requirements. These include using explosion-proof equipment and enclosures, as well as intrinsically safe devices that limit the energy released in a potential fault.
2. Healthcare facilities: Medical facilities have strict electrical safety requirements to protect patients and staff, particularly in areas where life-sustaining equipment is in use. Essential electrical systems must have adequate backup power, and isolated power systems are often required to minimize the risk of shock. Additionally, using ground fault circuit interrupters (GFCIs) in patient care areas can help prevent electrical hazards.
3. Outdoor installations: Electrical installations exposed to weather and other environmental conditions must be designed and installed to withstand these elements. This includes using weather-resistant materials, enclosures, and equipment, as well as providing GFCI protection for receptacles in wet or damp locations.

Understanding the unique safety requirements for different environments is vital for electrical workers. It ensures compliance with the NEC, reduces the risk of accidents, and promotes a safe working environment for everyone involved.

Safety around high voltage systems is crucial, as the risks associated with these systems are significantly higher than those of low voltage installations. High voltage electrical systems, typically defined as systems operating at voltages above 600 volts, demand specialized knowledge, training, and adherence to strict safety precautions. Here are some key safety practices to follow when working with high voltage systems:

1. Proper training and qualifications: Ensure that only qualified and trained professionals work on high voltage systems. This includes obtaining appropriate certifications, licenses, and completing specialized high voltage training courses.
2. Understand and follow safety regulations: Familiarize yourself with industry standards and regulations, such as the NEC, OSHA guidelines, and NFPA 70E, which outline specific safety requirements for high voltage systems.
3. Use appropriate personal protective equipment (PPE): Wear PPE designed for high voltage work, such as insulated gloves, voltage-rated footwear, flame-resistant clothing, and face shields.
4. Implement lockout/tagout procedures: Before working on high voltage equipment, ensure that the energy source is isolated, locked out, and tagged to prevent accidental energization.
5. Maintain a safe working distance: Keep a safe distance from energized high voltage equipment, and use insulated tools and equipment rated for the voltage level you're working with. Be aware of the approach boundaries defined by NFPA 70E.
6. Perform regular inspections and maintenance: Regularly inspect and maintain high voltage equipment to ensure its safe operation. Identify and address potential issues, such as damaged insulation or loose connections, before they become hazards.
7. Utilize appropriate grounding and bonding techniques: Proper grounding and bonding practices are essential for high voltage systems to ensure a safe path for fault currents and reduce the risk of electrical shock.
8. Test for voltage presence: Use voltage detectors and other testing equipment to verify the absence of voltage before working on any high voltage system.
9. Encourage a safety-oriented culture: Foster a work environment where safety is a top priority, and employees feel empowered to voice concerns and report potential hazards.

By following these safety practices and staying up-to-date with industry standards, electrical professionals can mitigate the risks associated with working on high voltage systems and ensure a safe working environment.

# **Blueprint Reading and Project Management**

Blueprint Reading and Project Management are essential skills for electrical professionals, particularly for those in leadership roles or working on complex installations. This chapter provides a comprehensive introduction to the principles of blueprint reading and project management, equipping you with the knowledge and tools necessary to interpret electrical drawings and effectively manage electrical projects from inception to completion.

In the Blueprint Reading section, you'll learn the basics of interpreting electrical blueprints, including the various symbols, notations, and types of drawings used in the industry. You'll become familiar with the purpose and structure of electrical schematics, wiring diagrams, and panel schedules, as well as how to navigate and comprehend complex electrical layouts. This knowledge will enable you to plan electrical installations, troubleshoot issues, and communicate effectively with colleagues and clients.

The Project Management section delves into the fundamentals of organizing, planning, and executing electrical projects. You'll learn about the stages of project management, including initiation, planning, execution, monitoring and control, and closure. Key concepts, such as scope definition, budgeting, scheduling, and risk management, will be explored to help you effectively manage resources, timelines, and stakeholder expectations. This section will also cover essential communication and leadership skills, empowering you to build a high-performing team and navigate the challenges of electrical projects successfully.

By mastering the skills of blueprint reading and project management, you'll be better equipped to handle complex electrical installations and ensure projects are completed safely, efficiently, and in compliance with industry standards and regulations.

Blueprint reading is a critical skill for electrical professionals, as it allows them to understand, interpret, and execute electrical installations and modifications in various settings. Electrical drawings, or blueprints, are visual representations of electrical systems, circuits, and components. They serve as a guide for electricians, engineers, and other professionals involved in the design, installation, and maintenance of electrical systems.
There are several types of electrical drawings, each with its specific purpose and focus. Here are some of the most common types:
1. Floor plans: These drawings provide a bird's-eye view of a building, showing the layout of rooms, walls, doors, windows, and electrical components such as outlets, switches, and lighting fixtures. Floor plans help electricians determine the best locations for wiring, devices, and equipment.
2. Schematics: Schematic diagrams represent electrical circuits using symbols and lines to depict components and connections. These drawings provide a detailed view of how individual components are connected and interact with each other, allowing electricians to troubleshoot and repair issues.

3. Wiring diagrams: Wiring diagrams show the physical connections and layout of an electrical system or circuit. They include information on wire sizes, colors, and routing, which helps electricians install and maintain the system accurately and safely.
4. Single-line diagrams: Single-line diagrams, or one-line diagrams, simplify complex electrical systems by representing them as single lines with symbols for various components. These diagrams are used to understand the overall organization of an electrical system and can be helpful for planning and troubleshooting.
5. Panel schedules: Panel schedules are documents that list the circuits and devices connected to an electrical panel. They include important information such as circuit numbers, breaker sizes, and loads, helping electricians manage and maintain the panel.

Mastering blueprint reading is essential for electrical professionals to ensure that installations are safe, code-compliant, and efficient. By understanding the various types of electrical drawings and their purposes, electricians can accurately plan, execute, and maintain electrical systems in a wide range of environments.

In electrical blueprints, various symbols, notations, and abbreviations are used to represent components and connections within the system. Being able to recognize and interpret these symbols is essential for understanding and working with electrical drawings. Here are some common symbols and notations you might encounter:

1. Wires and connections: Wires are typically represented by lines, with different line styles indicating different types of wires or connections. For example, a solid line might represent a single conductor, while a dashed line could indicate a multi-conductor cable.
2. Switches: Switches are represented by various symbols, depending on their type and function. A simple switch symbol consists of two lines with a break in the middle, while a three-way switch might have an additional line branching off.
3. Outlets and receptacles: Outlets are depicted by a small circle or rectangle, sometimes with a letter inside to indicate the outlet type (e.g., "G" for ground fault circuit interrupter or "D" for duplex receptacle).
4. Lighting fixtures: Lighting symbols can vary, but they generally include a circle or rectangle with an "L" or other letter inside to specify the fixture type (e.g., "FL" for fluorescent light).
5. Transformers: Transformers are represented by two or more coils, often with a core symbol connecting them. The number of windings on each coil may indicate the transformer's voltage ratio.
6. Circuit breakers and fuses: These protective devices are shown as rectangles or squares with a symbol or abbreviation to indicate their function (e.g., "CB" for circuit breaker or "F" for fuse).
7. Grounding: Grounding symbols include a downward-pointing triangle or a series of parallel lines decreasing in length, representing the connection to earth ground.

In addition to these symbols, electrical drawings may also use abbreviations to provide further information or to save space on the blueprint. Common abbreviations include:
- "V" for volts
- "A" for amperes
- "kVA" for kilovolt-amperes (transformer rating)
- "AWG" for American wire gauge (wire size)

- "R" for resistor
- "C" for capacitor
- "M" for motor

Understanding these symbols, notations, and abbreviations is crucial for interpreting electrical blueprints accurately and efficiently. Familiarizing yourself with these elements will enable you to read and work with electrical drawings confidently and ensure that your installations and repairs are safe and code-compliant.

Navigating electrical schematics is an essential skill for electrical professionals, as these diagrams provide a visual representation of the electrical system's layout, components, and connections. There are two main types of electrical schematics: single-line diagrams and multiline diagrams.

1. Single-line diagrams: These diagrams use a single line to represent multiple conductors, making them a simplified representation of an electrical system. They are typically used to illustrate the overall system layout, including the main components, such as generators, transformers, switchgear, and loads. Single-line diagrams provide a high-level view of the system, allowing you to understand the flow of power and the interrelationships between components.

To navigate a single-line diagram, start by identifying the power source(s) and follow the flow of electricity through the various components. Pay attention to symbols representing key components and devices, such as transformers, circuit breakers, and switches. Note any labels or numbering schemes used to identify specific components or connections, as these can help you locate the corresponding elements in other drawings or documentation.

2. Multiline diagrams: Unlike single-line diagrams, multiline diagrams use separate lines to represent each conductor in the system. These diagrams provide a more detailed view of the electrical system, showing individual wires, connections, and components. They are often used for troubleshooting and installation purposes, as they clearly depict the specific wiring connections required for each component.

To navigate a multiline diagram, begin by identifying the different conductors and their respective colors or labels. Follow the flow of electricity through the circuit, paying close attention to the various components, connections, and branching points. Take note of any symbols, abbreviations, or notations that may provide additional information about the components or connections. In some cases, multiline diagrams may include reference numbers or letters that correspond to specific devices or components on the drawing, helping you to identify the correct parts during installation or repair.

When working with electrical schematics, it's important to have a strong foundation in electrical symbols, notations, and abbreviations to ensure accurate interpretation. By familiarizing yourself with these elements and practicing your navigation skills on various types of diagrams, you'll be well-equipped to read and understand electrical schematics confidently and efficiently.

Wiring diagrams and panel schedules play a crucial role in electrical projects, as they provide essential information about the connections and organization of electrical components. These documents help ensure proper installation, maintenance, and troubleshooting of electrical systems.

1. Wiring diagrams: Wiring diagrams are detailed illustrations of the physical connections and layout of electrical components within a system. They provide a visual representation

of how devices and components are interconnected, illustrating the path of electrical wiring, along with any switches, receptacles, or other pertinent devices.

To read and use wiring diagrams effectively, begin by identifying the symbols representing various components, such as switches, outlets, or light fixtures. Follow the lines connecting these components to understand the flow of electricity through the system. Pay attention to any labels or color codes on the wires, as these can help you determine the purpose and function of each conductor. Additionally, take note of any notations or reference numbers that may provide further guidance during installation or troubleshooting.

2. Panel schedules: Panel schedules are organized lists that detail the various circuit breakers, fuses, or other protective devices within an electrical panel. They include information about the panel's capacity, the circuits being served, and the ratings of individual breakers or fuses. Panel schedules help to ensure proper load balancing and facilitate efficient maintenance and troubleshooting.

To read and use a panel schedule, start by examining the overall structure and layout of the document. Identify the panel's capacity and any associated subpanels, as well as the circuits being served by each breaker or fuse. Note the ratings of individual protective devices, which can help you determine if a circuit is overloaded or if additional capacity is available for future expansions. Additionally, pay attention to any labels, descriptions, or reference numbers that may provide further context about the circuits or devices within the panel.

In summary, wiring diagrams and panel schedules are vital tools in electrical projects, offering valuable information about the layout, connections, and organization of electrical components. By familiarizing yourself with these documents and learning to interpret them effectively, you'll be better equipped to ensure the successful completion and maintenance of your electrical projects.

Coordinating with other trades is essential when reading blueprints and planning electrical installations, as it ensures a smooth and efficient construction process. By collaborating with professionals from different fields, such as HVAC or plumbing, electrical workers can avoid potential conflicts, optimize space usage, and enhance overall system performance. Here are some key reasons why coordination is crucial:

1. Avoiding conflicts: Overlapping systems or equipment can create problems during installation or maintenance. By coordinating with other trades, electrical workers can identify potential issues early on and make the necessary adjustments to the design or installation plan, minimizing delays and costly rework.
2. Space optimization: Electrical components, HVAC equipment, and plumbing systems often share limited space within walls, ceilings, or floors. Proper coordination helps to ensure that each trade has adequate room to install and maintain their respective systems. This can lead to a more organized and efficient use of space, reducing clutter and facilitating access for future repairs or upgrades.
3. System integration: Modern buildings often feature integrated systems, where the performance of one component can affect others. For example, electrical systems may power HVAC equipment, which in turn can impact the energy consumption and comfort of the building. By coordinating with other trades, electrical workers can better understand the needs and requirements of these interconnected systems, leading to improved performance and energy efficiency.

4. Safety and compliance: Coordinating with other trades helps to ensure that all systems meet relevant safety codes and regulations. For instance, electrical workers must be aware of the required clearances around HVAC or plumbing components to prevent hazards such as fires or electrical shock. By working together, trades can guarantee that their installations are compliant and safe.
5. Improved communication: Coordination between trades fosters a collaborative environment where professionals can share knowledge, address concerns, and solve problems more effectively. This improved communication can lead to better overall project outcomes, as well as enhanced relationships among team members.

In conclusion, coordinating with other trades when reading blueprints and planning electrical installations is vital for a successful project. By collaborating with professionals from different fields, electrical workers can avoid conflicts, optimize space, enhance system performance, ensure safety and compliance, and improve communication.

Project management is a critical aspect of the electrical field, as it involves organizing, planning, executing, and controlling resources to achieve specific goals within a defined time frame and budget. Its significance lies in ensuring that electrical projects are completed efficiently, safely, and within the required quality standards. Here, we'll discuss an overview of project management in the electrical field and touch upon key concepts.

1. Planning: This phase entails defining the project scope, objectives, and deliverables. It involves setting a clear vision for the project, identifying required resources, and developing a detailed schedule. In the electrical field, planning often includes coordinating with other trades, assessing permits and regulations, and considering safety requirements.
2. Execution: During this stage, the actual work takes place, and the project team carries out the tasks outlined in the plan. For electrical projects, this may include installing wiring, setting up electrical panels, and connecting devices. Execution requires effective communication, problem-solving skills, and adaptability to address any unforeseen issues or changes in scope.
3. Monitoring and control: This phase involves tracking project progress, comparing it to the initial plan, and making necessary adjustments to keep the project on track. In the electrical field, monitoring may consist of regular site inspections, progress meetings, and quality assurance checks. Control mechanisms can include change management procedures and risk mitigation strategies.
4. Closure: Once the electrical work is completed, the project enters the closure phase. This involves ensuring all tasks are finalized, conducting a final review, and handing over the completed work to the client. Closure often includes a lessons-learned analysis to identify areas for improvement in future projects.
5. Key concepts in electrical project management:

a. Scope management: Defining and controlling the project's boundaries to ensure that only the necessary work is completed. b. Time management: Establishing a schedule and monitoring progress to ensure timely completion. c. Cost management: Estimating, budgeting, and controlling costs to deliver the project within the allocated budget. d. Quality management: Ensuring that the electrical work meets or exceeds the required standards and client expectations. e. Risk management: Identifying, analyzing, and addressing potential risks that could impact the

project's success. f. Communication management: Ensuring effective information exchange among project stakeholders.
In conclusion, project management in the electrical field is crucial for successfully completing projects on time, within budget, and with the desired quality. By understanding the various phases and key concepts, electrical professionals can better navigate complex projects and ensure successful outcomes.

Project management is a systematic approach to organizing, planning, executing, monitoring, controlling, and closing projects. It consists of different stages, each with specific objectives and tasks. Let's take a closer look at these stages and their key elements:

1. Initiation: This is the starting point of a project, where the idea is born, and its feasibility is assessed. The initiation stage involves identifying the project's purpose, establishing goals and objectives, and evaluating the project's viability in terms of cost, time, and resources. Key tasks include defining the project charter, identifying stakeholders, and conducting a feasibility study or a cost-benefit analysis.
2. Planning: The planning stage is where the project's roadmap is developed. It involves creating a detailed plan that outlines the scope, budget, schedule, resources, and other essential elements. This stage includes defining work breakdown structures, creating a project schedule with milestones, estimating costs and resources, and developing a risk management plan.
3. Execution: In the execution stage, the project plan is put into action. The project team works on completing the tasks outlined in the plan, while the project manager ensures that resources are allocated effectively, and progress is monitored. This stage may involve procuring materials, assigning tasks to team members, and conducting regular progress meetings to address any issues or obstacles that may arise.
4. Monitoring and control: This stage runs parallel to the execution phase and involves tracking the project's progress against the initial plan. The project manager monitors performance indicators, such as budget, schedule, and quality, and makes necessary adjustments to keep the project on track. Key tasks include updating the project schedule, controlling costs, managing risks, and ensuring quality standards are met.
5. Closure: The closure stage marks the end of the project. It involves wrapping up any remaining tasks, conducting a final review, and delivering the completed work to the client. This stage also includes evaluating the project's overall performance, identifying lessons learned, and documenting the project's successes and challenges for future reference.

In summary, the stages of project management—initiation, planning, execution, monitoring and control, and closure—provide a structured approach to managing projects. Each stage plays a critical role in ensuring a project's success by addressing various aspects of project development, from idea conception to final delivery. Understanding these stages enables project managers to effectively navigate complex projects and deliver the desired outcomes.

Defining Project Scope and Objectives:
Defining the project scope and setting clear objectives are crucial steps in ensuring a successful electrical project. The project scope outlines the boundaries of the project, including what is

included and what is excluded. Clear objectives help guide the team and provide a sense of direction throughout the project. Here's how to define project scope and set objectives:
1. Gather information: Consult with stakeholders, such as clients, team members, and other trade professionals, to gather their input and requirements. Review relevant regulations and standards, and assess site conditions to understand constraints and opportunities.
2. Define the scope: Clearly outline the project's boundaries, including the specific work to be performed, the deliverables, and any limitations or exclusions. Be as detailed as possible, as this will help prevent misunderstandings and scope creep later on.
3. Set SMART objectives: Objectives should be Specific, Measurable, Achievable, Relevant, and Time-bound. By setting SMART objectives, you ensure that the goals are clear, realistic, and can be easily tracked throughout the project.
4. Document and communicate: Document the project scope and objectives in a scope statement or project charter, and share it with all stakeholders. This ensures that everyone is on the same page and has a clear understanding of the project's expectations.

Budgeting and Resource Allocation:
Budgeting and resource allocation are essential aspects of project management, as they help ensure that a project is completed on time and within budget. Proper allocation of resources, such as labor, materials, and equipment, is crucial for a project's success. Here's how to approach budgeting and resource allocation:
1. Estimate costs: Begin by estimating the costs associated with each aspect of the project, such as labor, materials, equipment, permits, and other expenses. This will help you develop a comprehensive project budget.
2. Allocate resources: Based on the project's scope and objectives, allocate the necessary resources to each task or phase of the project. Consider factors such as skill levels, availability, and equipment requirements when allocating resources.
3. Monitor and adjust: Regularly monitor the project's progress, comparing actual costs and resource utilization with the initial budget and allocation plans. If necessary, adjust the allocation of resources or make other changes to stay within budget and meet project objectives.
4. Contingency planning: Include a contingency budget to account for unforeseen circumstances or changes in the project. This will provide a financial cushion and help prevent budget overruns.

By defining the project scope and objectives clearly, and effectively managing the budget and resource allocation, you can increase the likelihood of a successful electrical project, while minimizing risks and ensuring efficient use of resources.

Scheduling and Timeline Management:
Creating and managing project schedules is essential for ensuring that electrical projects are completed on time and within scope. Effective scheduling techniques include critical path analysis and Gantt charts. Here's an overview of these methods:
1. Critical Path Analysis (CPA): CPA is a technique used to determine the sequence of tasks that must be completed on time to finish the project as scheduled. It involves identifying the longest path of tasks (the critical path) and focusing on completing those tasks on time to avoid project delays. Steps for conducting CPA include: a. List all tasks: Identify all tasks required to complete the project. b. Estimate task durations: Estimate the time

needed to complete each task. c. Determine dependencies: Identify the relationships between tasks and which tasks must be completed before others can start. d. Calculate the critical path: Determine the longest path of dependent tasks and focus on completing these tasks on time.
  2. Gantt Charts: Gantt charts are visual tools that display the project schedule, showing the start and end dates of each task, their dependencies, and progress. They are useful for tracking the project's progress and making adjustments as needed. To create a Gantt chart: a. List tasks and durations: Identify all tasks and their estimated durations, as in the CPA. b. Define task dependencies: Determine which tasks depend on others and must be completed in sequence. c. Create the chart: Represent tasks as horizontal bars on a timeline, with the length of each bar corresponding to the task's duration. Show dependencies with connecting arrows.

Both CPA and Gantt charts can help you manage the project timeline effectively by highlighting critical tasks and dependencies, enabling you to make informed decisions and adjustments as needed.

Risk Management and Contingency Planning:

Risk management and contingency planning involve identifying, assessing, and mitigating risks in electrical projects. Here are strategies for managing risks in your project:
  1. Identify risks: Brainstorm potential risks, such as delays, cost overruns, accidents, or equipment failures. Consult with team members, stakeholders, and other trade professionals to gather their input.
  2. Assess risks: Evaluate the likelihood and potential impact of each identified risk. This will help you prioritize risks and allocate resources accordingly.
  3. Mitigate risks: Develop strategies to reduce the likelihood or impact of risks. This may involve implementing safety measures, improving communication, or adjusting the project schedule.
  4. Contingency planning: Create contingency plans for high-priority risks, outlining the steps to be taken in case a risk materializes. This may involve having backup equipment, alternative suppliers, or additional personnel available.
  5. Monitor and review: Regularly monitor the project's progress and reassess risks as the project evolves. Adjust risk mitigation strategies and contingency plans as needed.

By incorporating risk management and contingency planning into your project management process, you can minimize the potential impact of risks and increase the likelihood of a successful electrical project.

Communication and Leadership Skills:

Effective communication and leadership play a crucial role in managing electrical projects and building high-performing teams. Here's how these skills contribute to successful project management:
  1. Clear communication: Project managers must communicate project goals, expectations, and progress to their team members, stakeholders, and other trades involved in the project. Clear communication ensures everyone understands their roles and responsibilities, reducing the chances of confusion or misunderstandings.

2. Active listening: Active listening allows project managers to understand concerns, gather feedback, and address issues promptly. This skill is essential for building trust and fostering a collaborative work environment.
3. Decision-making: Project managers must make timely and informed decisions, balancing competing priorities and considering the potential impact on the project. Strong decision-making skills help maintain project momentum and prevent delays or cost overruns.
4. Delegation: Effective delegation involves assigning tasks to team members based on their skills and expertise, ensuring that work is completed efficiently and accurately. Delegation also promotes a sense of ownership and responsibility among team members.
5. Conflict resolution: Conflicts may arise during electrical projects due to differing opinions, miscommunications, or resource constraints. Project managers must be skilled at resolving conflicts in a constructive manner, maintaining positive relationships and fostering a collaborative work environment.

Quality Control and Assurance:
Quality control and assurance are essential for ensuring high-quality work in electrical projects. Here are some strategies for maintaining quality:
1. Establish quality standards: Clearly define the project's quality requirements, including performance specifications, industry standards, and regulatory compliance. Make sure all team members understand these standards.
2. Inspection and testing: Regularly inspect and test electrical work to ensure it meets quality standards and complies with applicable codes and regulations. Implement a system for documenting inspections and test results.
3. Training and certifications: Ensure team members have the necessary training, certifications, and experience to perform their tasks to the required quality standards.
4. Continuous improvement: Encourage a culture of continuous improvement by gathering feedback from team members, stakeholders, and clients. Use this feedback to identify areas for improvement and implement changes as needed.
5. Quality management system: Implement a quality management system (QMS) that provides a structured approach to managing quality in your projects. A QMS may include documented procedures, checklists, and performance metrics to track progress and ensure consistent quality throughout the project.

By emphasizing communication and leadership skills and prioritizing quality control and assurance, project managers can successfully manage electrical projects and deliver high-quality results that meet or exceed expectations.

Project closure is an essential step in the project management process, marking the completion of an electrical project. The process involves several key steps, including final inspections, documentation, and lessons learned.
1. Final inspections: At the end of an electrical project, a final inspection is carried out to ensure that all work meets the required quality standards and complies with applicable codes and regulations. This inspection typically involves verifying that the project has been completed according to the initial design, assessing the overall functionality of the electrical system, and confirming that all safety requirements have been met.
2. Documentation: Proper documentation is crucial for the successful closure of a project. This includes updating as-built drawings to reflect the final state of the electrical

installation, assembling operation and maintenance manuals for the client, and ensuring that all permits and approvals have been obtained. Additionally, the project manager should prepare a final report, summarizing the project's performance, budget, and timeline, as well as any deviations from the initial plan.
3. Lessons learned: After the project is completed, it's essential to gather the project team and stakeholders to discuss the project's successes, challenges, and areas for improvement. This process, often referred to as a "lessons learned" session, allows the team to reflect on their experiences and identify best practices that can be applied to future projects. It's important to document these lessons learned and incorporate them into the organization's knowledge base to foster continuous improvement and enhance the performance of future projects.
4. Final sign-off and handover: Once the project has passed the final inspection, and all documentation is complete, the project manager should obtain a final sign-off from the client, confirming that the project has met their expectations. The handover process includes transferring all relevant documents, manuals, and warranties to the client, as well as providing any necessary training on the operation and maintenance of the electrical system.

By following these steps, project managers can effectively close out electrical projects, ensuring that all work is completed to the highest standards and that valuable insights are captured for future projects.

# Electrical Calculations

Electrical calculations are a critical aspect of the electrical field, providing professionals with the necessary tools to design, install, and maintain safe, efficient, and code-compliant electrical systems. These calculations help ensure that electrical installations operate as intended, prevent potential hazards, and optimize energy consumption. For professionals such as electricians, engineers, and inspectors, a strong understanding of electrical calculations is essential for success in their careers.

In this chapter, we will explore a variety of key concepts related to electrical calculations, including:

1. Ohm's Law and power formulas: These fundamental equations relate voltage, current, resistance, and power in electrical circuits, providing the foundation for many electrical calculations.
2. Voltage drop calculations: To maintain proper voltage levels and prevent equipment damage, it is crucial to calculate and minimize voltage drop in electrical circuits.
3. Conductor sizing and ampacity: Selecting the appropriate size and type of conductors is critical for ensuring the safe and efficient operation of electrical installations.
4. Circuit protection and overcurrent devices: Proper selection and sizing of fuses and circuit breakers are essential for protecting electrical systems and equipment from damage due to overcurrent conditions.
5. Short circuit calculations: Understanding short circuit currents helps professionals design systems that can safely handle fault conditions.
6. Load calculations: Accurate load calculations for residential and commercial installations are essential for determining the capacity of electrical systems and ensuring they meet the demands of the connected devices.
7. Transformer, motor, and lighting calculations: These specialized calculations allow professionals to size and design transformers, motors, and lighting systems for various applications.
8. Energy consumption and efficiency calculations: Assessing and optimizing energy efficiency in electrical installations is an important aspect of sustainable design and can lead to cost savings and environmental benefits.

By mastering these concepts, electrical professionals will be better equipped to make informed decisions in their work, resulting in safer, more reliable, and energy-efficient electrical systems.

**Ohm's Law and power formulas** are fundamental equations in the electrical field, providing a foundation for numerous calculations. They help us understand the relationship between voltage, current, resistance, and power in electrical circuits, which is crucial for designing, installing, and maintaining electrical systems.

Ohm's Law states that the current (I) flowing through a conductor between two points is directly proportional to the voltage (V) across those points and inversely proportional to the resistance (R) of the conductor. Mathematically, it is represented as:

$I = V / R$

Where:
I = current (in amperes)
V = voltage (in volts)
R = resistance (in ohms)

Power formulas are used to calculate the power (P) consumed or generated in an electrical circuit. The power is the product of voltage and current. There are two main power formulas:

P = V * I
P = I^2 * R
Where:
P = power (in watts)
V = voltage (in volts)
I = current (in amperes)
R = resistance (in ohms)

Now, let's consider a real-world example to illustrate the application of Ohm's Law and power formulas. Suppose you have a 120V electrical circuit with a 60W light bulb connected to it. First, we'll use the power formula P = V * I to calculate the current flowing through the light bulb:

60W = 120V * I
I = 60W / 120V
I = 0.5A

Next, we can use Ohm's Law to find the resistance of the light bulb:

I = V / R
0.5A = 120V / R
R = 120V / 0.5A
R = 240 ohms

In this example, we used Ohm's Law and power formulas to determine the current flowing through the light bulb and its resistance. Understanding these equations allows electrical professionals to make informed decisions when designing and troubleshooting electrical systems, ensuring safety, reliability, and efficiency.

**Voltage drop** is an essential concept in electrical installations, as it refers to the decrease in voltage across a conductor due to the resistance encountered as current flows through it. A proper understanding of voltage drop is crucial for ensuring the efficient operation of electrical devices and preventing potential safety hazards, such as overheating or reduced performance.

The significance of voltage drop calculations lies in the need to maintain voltage levels within acceptable limits for the proper functioning of electrical devices. Voltage drop can be caused by various factors, including conductor length, conductor material, conductor cross-sectional area, and the amount of current flowing through the conductor.

There are different methods for calculating voltage drop, but one of the most commonly used formulas for a single-phase circuit is:

$$VD = (2 * L * R * I) / A$$

Where:
VD = voltage drop (in volts)
L = conductor length (in feet)
R = resistivity of the conductor material (in ohms per circular mil per foot)
I = current (in amperes)
A = conductor cross-sectional area (in circular mils)

For a three-phase circuit, the formula is slightly different:

$$VD = (\sqrt{3} * L * R * I) / A$$

It's important to note that the National Electrical Code (NEC) has specific recommendations for allowable voltage drop limits. For example, the NEC recommends a maximum voltage drop of 3% for branch circuits and 5% for feeder circuits.

Let's consider a real-world example to demonstrate the calculation of voltage drop. Suppose we have a single-phase circuit with a conductor length of 100 feet, a resistivity of 10 ohms per circular mil per foot, a current of 20 amperes, and a cross-sectional area of 10,380 circular mils (equivalent to a 6 AWG wire). Using the formula mentioned above, we can calculate the voltage drop as follows:

$$VD = (2 * 100ft * 10\Omega * 20A) / 10,380 \text{ cmil}$$
$$VD \approx 3.85V$$

In this example, the voltage drop across the conductor is approximately 3.85 volts. By understanding and calculating voltage drop, electrical professionals can make informed decisions when designing circuits, selecting appropriate conductor sizes, and ensuring the safe and efficient operation of electrical installations.

Selecting the appropriate size and type of conductors for electrical installations is crucial for ensuring safety, efficiency, and compliance with electrical codes. The process involves considering factors such as ampacity, temperature, and insulation. Let's discuss each factor and how it affects conductor sizing.
   1. Ampacity: Ampacity is the maximum current-carrying capacity of a conductor without exceeding its temperature rating. It is essential to select a conductor with an ampacity

equal to or greater than the circuit's expected maximum current to avoid overheating and potential fire hazards. The National Electrical Code (NEC) provides ampacity tables (such as Table 310.16) that list the allowable ampacities for different conductor sizes and insulation types.
2. Temperature: The temperature rating of a conductor's insulation has a significant impact on its ampacity. Conductors with higher temperature ratings can handle more current without overheating. When selecting a conductor, ensure that its temperature rating matches or exceeds the temperature requirements of the environment and the connected devices. The NEC also provides temperature correction factors for adjusting the ampacity of conductors when operating at temperatures other than their rated values.
3. Insulation: Different types of insulation materials have varying temperature ratings and, consequently, different ampacities. Common insulation types include THHN, THWN, XHHW, and RHW. Make sure to choose a conductor with insulation that meets the temperature and environmental requirements of the installation, such as resistance to moisture, chemicals, or physical damage.

When determining the appropriate conductor size for an electrical installation, follow these steps:
1. Calculate the expected maximum current for the circuit, considering factors such as load type, continuous or non-continuous operation, and any applicable demand factors.
2. Refer to the appropriate NEC ampacity table to find the minimum conductor size that can handle the calculated current, considering the insulation type and temperature rating.
3. If the installation environment has a temperature different from the conductor's rated temperature, apply the temperature correction factors provided in the NEC to adjust the conductor's ampacity accordingly.
4. For installations with multiple conductors in a conduit, apply adjustment factors to account for the increased heat generated by adjacent conductors.
5. Finally, consider any additional factors specific to the installation, such as voltage drop, physical constraints, or local code requirements, to select the most suitable conductor size and type.

By following these guidelines, electrical professionals can ensure the safe, efficient, and code-compliant installation of conductors in various electrical applications.

Circuit protection devices, such as fuses and circuit breakers, play a vital role in ensuring the safety and integrity of electrical systems. They are designed to interrupt the flow of current in a circuit when it exceeds a predetermined value, protecting components and wiring from damage due to overcurrent, short circuits, or ground faults. Let's discuss the purpose of these devices and provide guidance on selecting and sizing them for different applications.
1. Fuses: A fuse is a one-time-use device containing a thin conductive element that melts when the current exceeds its rated capacity. Fuses come in various sizes and types, including fast-acting and time-delay, to accommodate different applications. When selecting a fuse, consider factors such as the load type, maximum continuous current, and fault current. Choose a fuse with an ampere rating slightly higher than the circuit's maximum continuous current and ensure it can safely interrupt the fault current.
2. Circuit breakers: Circuit breakers are reusable devices that automatically disconnect a circuit when the current exceeds a specified value. They can be reset after a fault has been cleared, providing a more convenient and cost-effective solution compared to fuses. Circuit breakers come in various types, such as thermal, magnetic, and combination, each

with different trip characteristics. When selecting a circuit breaker, consider the load type, maximum continuous current, fault current, and any applicable code requirements.

To properly size circuit protection devices, follow these guidelines:
1. Determine the maximum continuous current of the circuit, taking into account factors such as load type, continuous or non-continuous operation, and any applicable demand factors.
2. Select a device with an ampere rating slightly higher than the calculated maximum continuous current to prevent nuisance tripping. The National Electrical Code (NEC) typically recommends sizing circuit breakers at 125% of the continuous load.
3. Ensure that the selected device can safely interrupt the fault current, as indicated by its interrupting rating or short-circuit current rating.
4. Consider any additional factors specific to the application, such as the need for selective coordination, arc-fault protection, or ground-fault protection, and choose the appropriate device accordingly.

By carefully selecting and sizing circuit protection devices, electrical professionals can provide essential protection against overcurrent, short circuits, and ground faults, safeguarding equipment and preventing electrical fires.

A short circuit is an unintended electrical connection between two points with differing voltage levels, resulting in an abrupt surge of current. This sudden flow of excessive current can lead to equipment damage, fires, and other hazards. Understanding and calculating short circuit currents is essential for electrical professionals to ensure proper equipment and circuit protection. Let's discuss the concept of short circuits and methods for calculating short circuit currents in various types of electrical systems.

There are two primary types of short circuits:
1. Phase-to-phase short circuit: This occurs when two or more energized conductors make an unintended connection. For example, when insulation fails between two adjacent phases in a three-phase system, a phase-to-phase short circuit occurs.
2. Phase-to-ground short circuit: This happens when an energized conductor makes an unintended connection with the ground or a grounded object. For example, when insulation fails between a phase conductor and the grounded metal enclosure of an electrical panel, a phase-to-ground short circuit occurs.

To calculate short circuit currents, you can use various methods, depending on the complexity and the type of the electrical system. Here are two common methods:
1. Point-to-point method: This is a simplified approach, often used for single-phase and balanced three-phase systems. It involves calculating the short circuit current at each point in the system by considering the source impedance, transformer impedance, and conductor impedance. The formula for short circuit current (Isc) is:

$Isc = V / Z\_total$

Where V is the voltage at the point of interest, and Z_total is the total impedance between the source and the point of interest.

2. Symmetrical components method: This method is used for unbalanced three-phase systems and considers the positive, negative, and zero-sequence components of the system. It requires more complex calculations and is typically performed using

specialized software. The method involves determining the sequence impedances at each point in the system and calculating the short circuit current based on these values.

By accurately calculating short circuit currents, electrical professionals can ensure the proper selection and sizing of circuit protection devices, such as fuses and circuit breakers, to protect electrical systems and equipment from the damaging effects of short circuits.

Load calculations are essential for designing safe and efficient electrical installations in both residential and commercial settings. These calculations help to determine the size of electrical equipment, conductors, and protection devices needed to handle the electrical load. Load calculations consider factors such as demand factors and diversity factors. Let's delve into these concepts and the methods used for calculating electrical loads.

Demand factors are ratios that represent the portion of the total connected load that is expected to be in use simultaneously. Demand factors account for the fact that not all electrical devices or appliances will be operating at the same time. They help to prevent overestimating the electrical load and ensure that electrical systems are not oversized.

Diversity factors represent the probability that multiple loads will operate at their maximum capacity simultaneously. A diversity factor greater than 1 indicates that the sum of individual maximum loads is greater than the actual maximum load for the group of loads. Considering diversity factors helps to design electrical systems that can handle actual load conditions more accurately.

For residential installations, load calculations typically involve the following steps:
1. Determine the general lighting load based on the square footage of the residence.
2. Calculate the appliance and fixed equipment load, considering the wattage of each device and the demand factors.
3. Identify the largest motor load in the residence, and calculate the additional load it contributes.
4. Sum up all loads calculated in the previous steps, and apply any applicable demand factors.

For commercial installations, load calculations may involve additional steps, as these installations usually have more complex and varied electrical loads. Some steps include:
1. Determine the general lighting and receptacle load for the commercial building.
2. Calculate the specific loads for equipment, such as motors, heating, ventilation, and air conditioning (HVAC) systems, and other specialized devices.
3. Apply demand and diversity factors to account for the simultaneous operation of multiple devices or systems.
4. Sum up all the loads, considering any applicable demand and diversity factors.

By performing accurate load calculations for residential and commercial installations, electrical professionals can ensure that electrical systems are designed to handle the anticipated load while maintaining safety and efficiency. This helps to prevent overloaded circuits, equipment damage, and potential hazards.

**Transformers are essential electrical devices used to step up or step down voltage levels in alternating current (AC) systems. They enable efficient power transmission and distribution by adjusting voltage levels according to the requirements of different applications. In this section,**

we'll discuss the principles of transformer operation and provide guidance on selecting and sizing transformers, including calculations for voltage, current, and power.

The operation of a transformer is based on the principle of electromagnetic induction. A transformer consists of two or more coils, called windings, which are wrapped around a magnetic core. When an AC voltage is applied to the primary winding, it creates a varying magnetic field around the core, which in turn induces an AC voltage in the secondary winding. The ratio of the number of turns in the primary and secondary windings determines the voltage transformation.

To select and size transformers for different applications, you need to consider the following parameters:

1. Voltage: The primary and secondary voltages (V1 and V2) of the transformer are related to the number of turns in the primary (N1) and secondary (N2) windings as follows:

$V2/V1 = N2/N1 = $ Turns Ratio

2. Current: The primary and secondary currents (I1 and I2) are inversely related to the voltage and turns ratio:

$I1/I2 = N2/N1 = $ Turns Ratio

3. Power: The power rating of a transformer is specified as the apparent power (in volt-amperes or VA) or the real power (in watts) that it can handle. The power rating should be equal on both the primary and secondary sides:

$V1 \times I1 = V2 \times I2$

To select a transformer, you should consider the application requirements, such as the input and output voltages, the load current, and the power rating. To size a transformer, follow these steps:

1. Determine the required primary and secondary voltages based on the application needs.
2. Calculate the turns ratio using the voltage values.
3. Estimate the load current and ensure that the transformer can handle the required current.
4. Choose a transformer with a power rating that meets or exceeds the calculated power requirement.

By understanding the principles of transformer operation and performing accurate calculations for voltage, current, and power, you can select and size transformers that are suitable for various applications, ensuring efficient and reliable operation in electrical systems.

Electric motors play a crucial role in converting electrical energy into mechanical energy in various applications. To select and size motors appropriately, it's essential to understand key parameters such as horsepower, torque, and efficiency. In this section, we'll explain these parameters and provide guidance on selecting and sizing motors, including calculations for motor current, voltage, and power.

1. Horsepower (HP): Horsepower is a unit of power that measures a motor's capacity to perform work. It represents the rate at which work is done or energy is transferred. One horsepower is equivalent to 746 watts.
2. Torque (T): Torque is the rotational force generated by a motor, typically measured in Newton-meters (Nm) or pound-feet (lb-ft). Torque is the force that causes an object to rotate around an axis.
3. Efficiency ($\eta$): Efficiency is the ratio of the mechanical output power to the electrical input power of a motor, usually expressed as a percentage. Higher efficiency motors convert more electrical energy into mechanical energy, resulting in lower energy consumption and operating costs.

To select and size motors for different applications, consider the following steps:
1. Determine the required horsepower based on the application's power needs. This may depend on factors such as the load type, operating speed, and duty cycle.
2. Calculate the torque needed for the application using the formula:

$T = (HP \times 5252) / RPM$

where T is torque (in lb-ft), HP is horsepower, and RPM is the motor speed in revolutions per minute.
3. Choose a motor with a suitable efficiency rating, taking into account energy consumption and operating costs.
4. Calculate the full-load current (FLC) of the motor using the formula:

$FLC (Amps) = (HP \times 746) / (V \times \eta \times PF)$

where FLC is full-load current (in amps), HP is horsepower, V is the motor's voltage, $\eta$ is efficiency, and PF is the power factor.
5. Select a motor with a voltage rating that matches the available supply voltage.
6. Ensure that the motor's power rating (in watts or horsepower) meets or exceeds the calculated power requirement.

By understanding the key parameters of electric motors and performing accurate calculations for horsepower, torque, efficiency, and other factors, you can select and size motors that are suitable for various applications, ensuring efficient and reliable performance.

Designing and sizing lighting systems involves calculating illuminance levels, determining the number of fixtures needed, and estimating energy consumption. This process ensures that spaces are adequately lit while also considering energy efficiency and cost-effectiveness.
1. Illuminance levels: Illuminance is the amount of light incident on a surface, measured in foot-candles (fc) or lux (lx). To calculate illuminance levels, you need to consider the purpose of the space, task visibility, and any applicable codes or standards. For example, office spaces typically require higher illuminance levels than hallways or storage areas.
2. Fixture quantities: To determine the number of fixtures needed, first calculate the total lumens required for the space. This is done using the formula:

Total Lumens = Area (sq.ft or sq.m) × Desired Illuminance (fc or lx)

Next, divide the total lumens by the lumens per fixture to find the required number of fixtures:

Number of Fixtures = Total Lumens / Lumens per Fixture

Keep in mind that these calculations assume uniform distribution of light. In practice, you may need to adjust the number of fixtures to account for factors such as fixture spacing, mounting height, and light distribution patterns.
3. Energy consumption: To estimate energy consumption, you need to consider the wattage of the chosen fixtures and their operating hours. Calculate the total energy consumption using the formula:

Energy Consumption (kWh) = Number of Fixtures × Fixture Wattage (W) × Operating Hours / 1000

This information can help you select energy-efficient lighting solutions and estimate operating costs.

When designing and sizing lighting systems, consider factors such as illuminance levels, fixture quantities, and energy consumption. By performing accurate calculations and adhering to applicable codes and standards, you can create lighting systems that provide adequate illumination while also promoting energy efficiency and cost-effectiveness.

Energy efficiency is crucial in electrical installations because it helps reduce energy costs, decrease greenhouse gas emissions, and lower the strain on electrical grids. By calculating energy consumption and implementing strategies to improve efficiency, you can optimize the performance of electrical systems and minimize their environmental impact.

Methods for calculating energy consumption:
1. Determine the power consumption of individual devices or systems by multiplying their voltage (V) and current (A) to find the power (P) in watts (W). For example, $P = V \times A$.
2. Calculate the energy consumption by multiplying the power consumption (W) by the number of operating hours. Then, divide by 1000 to obtain energy consumption in kilowatt-hours (kWh). For example, Energy Consumption (kWh) = (Power Consumption x Operating Hours) / 1000.

Strategies for improving efficiency:
1. Energy-efficient devices: Select energy-efficient appliances, lighting, and equipment that consume less energy while maintaining high performance levels. Look for Energy Star-rated products or other efficiency certifications to ensure energy savings.
2. Proper sizing of equipment: Oversized or undersized equipment can lead to inefficiencies and higher energy consumption. Carefully assess the requirements of your electrical installations and choose appropriately sized equipment to optimize energy use.
3. Regular maintenance: Proper maintenance of electrical systems, including cleaning, inspecting, and replacing worn components, can help maintain optimal performance and reduce energy waste.
4. Lighting controls: Incorporate occupancy sensors, timers, or daylight sensors to control lighting and reduce energy consumption when spaces are unoccupied or when natural light is available.
5. Power factor correction: Improve the power factor of electrical systems by installing capacitors or other corrective devices. This can help reduce energy losses and improve overall efficiency.
6. Variable frequency drives (VFDs): Install VFDs for motors in applications where variable speeds are needed. VFDs can optimize motor performance, resulting in energy savings.
7. Insulation and sealing: Proper insulation and sealing of buildings can minimize heat loss or gain, reducing the energy demand on HVAC systems.

Understanding the importance of energy efficiency and implementing strategies to calculate and improve energy consumption can lead to optimized electrical installations, cost savings, and reduced environmental impact.


Problem 1: Ohm's Law

A 60W light bulb is connected to a 120V power source. What is the current flowing through the bulb?

Step 1: Write down the given information and the formula.

Power (P) = 60W
Voltage (V) = 120V
Current (I) = ?

Ohm's law formula: P = V x I

Step 2: Solve for the current.

I = P / V
I = 60W / 120V
I = 0.5A

Answer: The current flowing through the bulb is 0.5A.

Problem 2: Voltage Drop Calculation

A 200-ft long 12 AWG copper wire carries a 20A current. The wire has a resistance of 1.588 ohms per 1000 feet. Calculate the voltage drop across the wire.

Step 1: Write down the given information and the formula.

Current (I) = 20A
Resistance per 1000 ft (R) = 1.588 ohms
Wire length (L) = 200 ft

Voltage drop formula: V_drop = I x R x (L / 1000)

Step 2: Calculate the resistance for the 200-ft wire.

R_200 = 1.588 ohms * (200 ft / 1000 ft)
R_200 = 0.3176 ohms

Step 3: Calculate the voltage drop.

V_drop = 20A * 0.3176 ohms
V_drop = 6.352V

Answer: The voltage drop across the 200-ft wire is 6.352V.

Problem 3: Transformer Sizing

A single-phase transformer is required to supply a load of 50 kVA with a primary voltage of 480V and a secondary voltage of 240V. What is the current on the primary and secondary sides of the transformer?

Step 1: Write down the given information and the formula.

Apparent power (S) = 50 kVA
Primary voltage (V_primary) = 480V
Secondary voltage (V_secondary) = 240V
Primary current (I_primary) = ?
Secondary current (I_secondary) = ?

Transformer formula: S = V_primary x I_primary = V_secondary x I_secondary

Step 2: Solve for primary current.

I_primary = S / V_primary
I_primary = 50,000VA / 480V
I_primary = 104.17A

Step 3: Solve for secondary current.

I_secondary = S / V_secondary
I_secondary = 50,000VA / 240V
I_secondary = 208.33A

Answer: The primary current is 104.17A, and the secondary current is 208.33A.

These problems cover a range of electrical calculation topics, including Ohm's law, voltage drop calculations, and transformer sizing. Practice these types of problems to develop your skills and understanding of electrical calculations.

# **Installation and Troubleshooting**

Selecting and installing electrical panels is an essential step in the process of designing and building electrical systems. The panel serves as the central point for distributing power throughout a building and houses circuit protection devices, such as circuit breakers or fuses. When selecting and installing electrical panels, consider the following factors: panel size, location, and safety considerations.
1. Panel size: The size of the electrical panel is determined by the number of circuits it needs to accommodate, as well as the total electrical load in the building. When selecting a panel, consider the number of circuits required for the installation, as well as any potential future expansions. Panels come in different sizes, ranging from small units with just a few circuits to large industrial panels with numerous circuit breakers.
2. Location: The location of the electrical panel is crucial for both convenience and safety. Panels should be installed in easily accessible locations, such as utility rooms, basements, or garages. Avoid locations that are exposed to moisture or excessive heat, as these can compromise the panel's performance and safety. Additionally, ensure that the panel is installed in a location with enough clearance around it to allow for easy maintenance and safe operation. National and local electrical codes will provide specific guidelines for panel location and clearance requirements.
3. Safety considerations: Safety is of utmost importance when installing electrical panels. Follow these safety guidelines to ensure a safe and effective installation:
- Always adhere to national and local electrical codes, as well as manufacturer instructions for panel installation.
- Turn off power to the area where the panel will be installed before beginning work. Use lockout/tagout procedures if necessary.
- Use the proper personal protective equipment (PPE) when working with electrical systems, including gloves, safety glasses, and insulated tools.
- Make sure that the panel is properly grounded and bonded to prevent electrical shock hazards.
- Label all circuit breakers or fuses with their respective loads to facilitate easy identification and maintenance.
- Install arc fault circuit interrupters (AFCIs) and ground fault circuit interrupters (GFCIs) as required by the electrical code to protect against electrical hazards.

By carefully considering panel size, location, and safety guidelines during the installation process, you can ensure that your electrical panel is efficient, functional, and safe.

Wiring methods and techniques are essential in the electrical installations process. These methods involve the use of conduits, various cable types, and wire connections to create safe and efficient electrical systems. Let's delve into each of these aspects:
1. Conduit: Conduit is a protective tubing used to house electrical wires, providing both physical protection and a secure routing path. There are several types of conduits, such as:

- Rigid metal conduit (RMC): This strong and durable conduit is made of steel or aluminum and provides excellent protection against physical damage.
- Intermediate metal conduit (IMC): IMC is lighter than RMC, but still offers good protection.
- Electrical metallic tubing (EMT): EMT is a thin-walled, lightweight option that's easy to work with, but offers less protection than RMC or IMC.
- Flexible metal conduit (FMC): FMC, also known as Greenfield, is a flexible option used in tight spaces or when routing around obstacles.
- Non-metallic conduits: These conduits, such as PVC or liquid-tight flexible nonmetallic conduit (LFNC), offer corrosion resistance and are suitable for damp or corrosive environments.

2. Cable types: Several cable types are used in electrical installations, each designed for specific purposes and applications:
- Nonmetallic sheathed cable (NM cable): Also known as Romex, this cable is commonly used in residential wiring due to its ease of installation and lower cost.
- Armored cable (AC): This type of cable includes a flexible metal armor that protects the conductors. It's suitable for commercial and industrial applications.
- Metal-clad cable (MC): MC cable is similar to AC cable but offers even more protection, with a grounding conductor included.
- Underground feeder (UF) cable: UF cable is designed for direct burial or use in wet locations, featuring a moisture-resistant construction.

3. Wire connections: Proper wire connections are crucial for safety and efficiency. There are several techniques and devices used to make secure connections:
- Wire nuts: These are used to join two or more wires together by twisting them around each other and securing them with a wire nut, which provides insulation and protection.
- Terminal blocks and screw terminals: These devices allow wires to be connected securely by clamping them under a screw or pressure plate.
- Crimp connections: A crimping tool is used to compress a metal sleeve around the stripped ends of two or more wires, creating a strong and secure connection.
- Soldering: This technique involves melting a metal alloy (solder) to join wires together, creating a permanent and reliable connection. Soldering is less common in modern electrical installations due to the time and skill required.

By understanding and implementing these wiring methods and techniques, you can ensure that your electrical installations are safe, efficient, and compliant with applicable codes and standards.


Grounding and bonding are crucial components of electrical installations, providing safety, stability, and fault protection. Let's discuss the importance of grounding and bonding and explore the methods for implementing them in electrical systems.

Importance of grounding and bonding:

1. Safety: Grounding and bonding help protect people from electric shock by creating a low-impedance path for fault currents to flow, reducing the potential for hazardous voltages on accessible conductive surfaces.
2. Fault protection: A properly grounded and bonded system facilitates the operation of overcurrent protection devices, such as fuses or circuit breakers, by allowing fault currents to flow safely to the ground, triggering the protective devices to disconnect the faulted circuit.
3. Voltage stabilization: Grounding helps stabilize the voltage in the electrical system by providing a reference point connected to the earth, reducing the risk of voltage fluctuations that can damage equipment.
4. Lightning and surge protection: Grounding and bonding help dissipate the energy from lightning strikes or surges, protecting the electrical system and connected equipment.

Methods for grounding and bonding:
1. System grounding: Connect the electrical system's neutral point (usually the neutral bus in the main service panel) to a grounding electrode, such as a ground rod or metallic water pipe. This connection creates an intentional low-impedance path for fault currents to flow to the earth.
2. Equipment grounding: Establish a low-impedance path from all non-current-carrying metallic parts of the electrical system, such as enclosures and conduits, back to the main grounding point. This path is typically established using equipment grounding conductors (EGCs) that run alongside the current-carrying conductors.
3. Bonding: Electrically connect all metallic components of the electrical system to create a continuous, low-impedance path. Bonding can be achieved through methods such as using bonding jumpers, exothermic welding, or mechanical clamps. The main bonding jumper connects the grounded (neutral) conductor to the equipment grounding conductors at the main service panel.
4. Grounding electrodes: Install grounding electrodes, such as ground rods, metal underground water pipes, or concrete-encased electrodes, to create a direct connection to the earth. The grounding electrode conductor (GEC) connects the grounding electrode(s) to the main grounding point.

By implementing proper grounding and bonding techniques in electrical installations, you can ensure a safe and stable electrical system that minimizes the risk of electric shock, equipment damage, and fires caused by electrical faults.

Installing electrical devices like receptacles, switches, and light fixtures is a common task in electrical installations. The following steps will guide you through the process while emphasizing safety precautions and best practices.
1. Safety first: Before starting, ensure the power is turned off at the circuit breaker or fuse box. Use a voltage tester to confirm that there's no live voltage present at the installation location. Always wear appropriate personal protective equipment (PPE), such as gloves and safety glasses.

2. Select the right device: Choose electrical devices that are compatible with your electrical system and meet the requirements of the application. For example, use GFCI receptacles in wet locations, or select switches with the appropriate current rating.
3. Prepare the device box: Install an appropriate device box at the installation location, following local codes and manufacturer recommendations. The box should be securely fastened and properly aligned with the wall surface.
4. Run the wiring: Route the necessary wiring to the device box, ensuring you follow local codes and best practices for cable installation. Use appropriate cable clamps and supports to secure the wiring.
5. Strip and connect wires: Strip the insulation from the ends of the wires according to the device's requirements. Use wire connectors or terminal screws to make secure connections between the device and the wires. Make sure to connect the wires according to their functions: hot (usually black or red) to the brass terminal, neutral (white) to the silver terminal, and the ground (green or bare) to the green terminal or grounding screw.
6. Attach the device: Carefully fold the wires into the device box, ensuring that they are not pinched or strained. Fasten the device to the box using the provided mounting screws, ensuring the device is level and firmly secured.
7. Install cover plates: Attach the appropriate cover plate to the device, making sure it is aligned and securely fastened.
8. Test the installation: Restore power to the circuit and test the installed device to ensure proper operation. Use a plug-in tester for receptacles, and verify that switches and light fixtures operate as intended.

By following these steps and adhering to safety precautions, you can successfully install common electrical devices in a professional and reliable manner. Keep in mind that local codes and regulations may vary, so always consult with your local authority having jurisdiction (AHJ) to ensure compliance with any specific requirements.

Circuit protection devices, such as fuses and circuit breakers, are crucial for safeguarding electrical installations from overcurrents and potential damage. Proper installation and configuration of these devices are vital to ensure safety and reliability. Here's a guide on how to install and configure fuses and circuit breakers, considering factors like amperage, voltage, and interruption ratings.

1. Determine the right device: Start by selecting the appropriate circuit protection device based on the application and electrical system. Consider the required amperage, voltage, and interruption ratings to ensure the device can safely handle the load and potential fault currents.
2. Calculate the required rating: Perform load calculations to determine the required amperage rating for the device. In general, the rating should be higher than the continuous load current but lower than the maximum allowable current for the wiring.

Follow the National Electrical Code (NEC) or local regulations to calculate the right rating.
3. Choose the right voltage rating: Select a device with a voltage rating that matches or exceeds the system voltage. The voltage rating indicates the maximum voltage the device can safely interrupt in case of a fault.
4. Select the appropriate interruption rating: The interruption rating, expressed in amperes, represents the maximum fault current the device can safely interrupt. Choose a device with an interruption rating higher than the prospective fault current at the point of installation.
5. Turn off the power: Before installing the circuit protection device, switch off the power at the main disconnect and verify that the area is de-energized using a voltage tester.
6. Install the device: For fuses, insert the fuse into the fuse holder, ensuring proper alignment and a snug fit. For circuit breakers, snap the breaker into the panel's busbar and secure it according to the manufacturer's instructions.
7. Wire the device: Connect the hot wire (usually black or red) to the device's terminal, ensuring a secure connection. The neutral (white) and ground (green or bare) wires should be connected to the neutral and grounding bars in the panel, respectively.
8. Verify the installation: Double-check that the device is installed correctly and securely, and that all connections are tight and properly positioned.
9. Turn the power back on: Restore power at the main disconnect, then switch on the newly installed circuit protection device.
10. Test the installation: Confirm that the protected circuit is functioning correctly, and check the device's operation by simulating an overcurrent situation or using a suitable testing device.

By following these steps and adhering to safety precautions, you can effectively install and configure circuit protection devices in various applications. Keep in mind that local codes and regulations may vary, so always consult your local authority having jurisdiction (AHJ) to ensure compliance with any specific requirements.


Troubleshooting electrical circuits is an essential skill for electricians and professionals working with electrical systems. A systematic approach can help you identify and resolve issues effectively and safely. Here's a guide on troubleshooting electrical circuits, including using test equipment such as multimeters and clamp meters.
1. Safety first: Before starting any troubleshooting, ensure you follow safety protocols. Turn off the power to the circuit, use appropriate personal protective equipment (PPE), and work in a well-lit environment.
2. Gather information: Collect as much information as possible about the problem, including any symptoms, recent changes, or modifications to the system.
3. Visual inspection: Perform a thorough visual inspection of the circuit, looking for obvious issues such as damaged components, loose connections, or signs of overheating.

4. Develop a hypothesis: Based on your observations and gathered information, create a hypothesis for the cause of the problem.
5. Test your hypothesis: Use test equipment, such as multimeters and clamp meters, to verify your hypothesis. Here's how to use these tools:

a. Multimeter: A multimeter is a versatile tool that measures voltage, current, and resistance. Set the multimeter to the appropriate function (AC or DC voltage, current, or resistance), and connect the test leads to the circuit component you want to test. Read the measurement displayed on the multimeter's screen.

b. Clamp meter: A clamp meter is used to measure current without disconnecting the circuit. Open the clamp, place it around a single conductor, and close the clamp. The meter will display the current flowing through the conductor.

6. Isolate the issue: If the test results confirm your hypothesis, proceed to isolate the faulty component or connection. If not, reevaluate your hypothesis and continue testing until the issue is identified.
7. Repair or replace: Once you've identified the problem, repair or replace the faulty component or connection as needed.
8. Verify the solution: After repairing or replacing the component, restore power to the circuit and verify that the issue has been resolved.
9. Document your findings: Keep a record of the troubleshooting process, the identified issue, and the solution implemented. This documentation can be helpful for future reference and maintenance.

By following this systematic approach and using test equipment like multimeters and clamp meters, you can efficiently and safely troubleshoot electrical circuits. Remember to always prioritize safety and consult with experienced professionals or reference materials when needed.


Common electrical problems encountered in residential and commercial installations can be broadly categorized into open circuits, short circuits, and overloads. Here's a discussion of each issue and guidance on how to repair them:

1. Open circuits: An open circuit occurs when there's a break in the electrical path, preventing current from flowing. Common causes include damaged wires, loose connections, or faulty components such as switches or receptacles.

Repair: a. Turn off the power to the affected circuit and use a multimeter to test for continuity. b. Inspect the wiring and components for any visible damage or loose connections. c. Repair or replace damaged wires, tighten loose connections, or replace faulty components as necessary. d. Restore power and test the circuit to ensure the issue is resolved.

2. Short circuits: A short circuit happens when a low-resistance path allows excessive current to flow, bypassing the intended load. This can lead to overheating, component damage, and even fire. Common causes include damaged insulation, pinched wires, or faulty devices.

Repair: a. Turn off the power and use a multimeter to check for low resistance between conductors or between a conductor and ground. b. Visually inspect the circuit for damaged insulation, pinched wires, or signs of overheating. c. Repair or replace any damaged wires, separate pinched wires, or replace faulty devices as needed. d. Restore power and test the circuit to ensure the short circuit has been eliminated.

3. Overloads: An overload occurs when the electrical load exceeds the capacity of a circuit, causing excessive heat and potential damage. Common causes include too many devices connected to a single circuit, faulty devices drawing excessive current, or incorrectly sized circuit components.

Repair: a. Identify the cause of the overload, such as an excessive number of devices or a faulty device drawing too much current. b. Unplug or disconnect unnecessary devices, replace faulty devices, or redistribute the load across multiple circuits. c. Ensure that circuit components such as wires, breakers, or fuses are correctly sized for the load they're serving. d. If needed, consult an electrician to upgrade the circuit capacity or add new circuits to handle the load safely.

When addressing these common electrical problems, always prioritize safety and adhere to local electrical codes. If you're unsure about any aspect of the repair process, consult with a qualified electrician to ensure a safe and effective resolution.

The importance of safety practices in electrical work cannot be overstated, as it helps prevent accidents, injuries, and even fatalities. Adhering to safety regulations and using personal protective equipment (PPE) are essential aspects of maintaining a safe work environment.

1. Personal Protective Equipment (PPE): Using appropriate PPE is crucial to protect yourself from potential hazards while working with electricity. Some common PPE items include:

a. Insulated gloves: Protect your hands from electrical shock and burns. b. Safety glasses: Shield your eyes from debris, sparks, or arc flashes. c. Insulated tools: Prevent electrical conduction, reducing the risk of shock. d. Flame-resistant clothing: Minimize the risk of severe burn injuries from arc flashes. e. Voltage-rated shoes: Offer additional protection against electrical shock.

2. Adherence to safety regulations: Following safety regulations and guidelines ensures the well-being of all workers involved in electrical installations and troubleshooting. Some key safety practices include:

a. Lockout/tagout procedures: Isolate and secure energy sources during maintenance or repair work to prevent accidental re-energizing of the circuit. b. Proper grounding and bonding: Minimize the risk of electrical shock and prevent damage from voltage fluctuations or lightning strikes. c. Regular equipment maintenance: Inspect and maintain electrical systems regularly to identify and address potential hazards. d. Training and qualifications: Ensure that personnel working on electrical installations and troubleshooting have the necessary skills and qualifications. e. Hazard assessment: Identify and evaluate potential hazards before starting work, and implement appropriate control measures.

By prioritizing safety practices and understanding the importance of PPE and adherence to safety regulations, you can significantly reduce the risk of accidents and injuries in electrical work. Remember to stay informed about updates to safety regulations and always follow the guidelines set forth by your local regulatory authority.

Preventive maintenance plays a crucial role in maintaining the reliability, safety, and efficiency of electrical systems. By performing regular maintenance tasks, you can identify potential issues before they become major problems, extend the lifespan of equipment, and minimize the risk of unexpected failures and costly downtime.

Common preventive maintenance tasks for electrical systems include:

1. Inspecting connections: Regularly examine all electrical connections in panels, switches, receptacles, and other devices to ensure they are secure and free of signs of overheating or corrosion. Loose or damaged connections can lead to increased resistance, overheating, and even electrical fires.
2. Testing devices: Periodically test devices such as circuit breakers, fuses, and ground fault circuit interrupters (GFCIs) to ensure they are functioning correctly. This can help detect any potential malfunctions early on and ensure the devices provide proper protection in case of a fault.
3. Checking wire insulation: Inspect the insulation of wires and cables for any signs of damage, such as cracks, tears, or discoloration. Damaged insulation can expose conductors and increase the risk of electrical shock or short circuits.
4. Cleaning and dusting: Remove dust and debris from electrical equipment, such as panels and enclosures, to prevent overheating and improve the overall performance of the system. Use appropriate cleaning tools and methods to avoid damaging any components.
5. Lubricating moving parts: Apply lubrication to moving parts in equipment like motors, switches, and circuit breakers to minimize wear and extend their service life.
6. Thermal imaging: Use infrared thermography to detect hotspots in electrical systems, which could indicate potential issues such as loose connections, overloaded circuits, or failing components.
7. Recordkeeping: Maintain accurate records of all maintenance tasks performed, including dates, findings, and any corrective actions taken. This information can help track the performance of the electrical system over time and identify trends or recurring issues.

By incorporating preventive maintenance into your regular routine, you can significantly improve the safety, reliability, and efficiency of your electrical systems. This proactive approach can also lead to cost savings by reducing the need for reactive maintenance and minimizing the impact of unexpected equipment failures.

Troubleshooting and replacing transformers is an essential skill for electrical professionals, as transformers play a critical role in a wide range of applications, from power distribution to control circuits. The process involves identifying failed transformers, understanding the cause of the failure, and selecting appropriate replacements.

1. Identifying failed transformers: Signs of a failed transformer can include erratic voltage readings, abnormal noises, overheating, or visible signs of damage, such as leaks, bulging, or scorch marks. To confirm a transformer failure, use a multimeter to measure the input and output voltages, and compare them to the manufacturer's specifications. If there's a significant discrepancy, the transformer might be faulty.
2. Determining the cause of failure: It's important to understand why the transformer failed to prevent similar issues in the future. Common causes of transformer failure include:
    - Overloading: Exceeding the transformer's rated capacity can cause overheating and eventual failure. Ensure the connected load doesn't exceed the transformer's specifications.
    - Short circuits: A short circuit in the secondary winding can cause excessive current flow, leading to overheating and damage.
    - Environmental factors: Exposure to moisture, dust, or corrosive environments can degrade transformer insulation and cause failure.
    - Manufacturing defects: Occasionally, manufacturing defects can lead to premature failure.
3. Selecting the appropriate replacement: When selecting a replacement transformer, consider the following factors:
    - Voltage ratings: Match the primary and secondary voltage ratings of the original transformer.
    - Power rating: Ensure the replacement transformer's power rating (measured in VA or kVA) is equal to or greater than the original.
    - Impedance: Select a transformer with a similar impedance to maintain compatibility with the existing electrical system.
    - Physical dimensions: Choose a replacement transformer with similar dimensions to ensure it fits in the available space.
    - Cooling method: Verify that the cooling method of the replacement transformer is appropriate for the application and environment.
4. Replacing the transformer: Once you have selected the appropriate replacement, follow these general steps:
    - Disconnect power: Ensure all power sources are disconnected and locked out before working on the transformer.
    - Remove the old transformer: Disconnect all wiring connections, and carefully remove the failed transformer from its mounting.
    - Install the new transformer: Secure the replacement transformer in place, following the manufacturer's instructions and adhering to any applicable codes and standards.
    - Connect wiring: Reconnect the primary and secondary wiring according to the manufacturer's guidelines, taking care to maintain proper polarity.

- Test the new transformer: Power up the system and use a multimeter to confirm that the input and output voltages match the manufacturer's specifications.

By following these steps, you can effectively troubleshoot and replace transformers in various applications, ensuring the safe and efficient operation of your electrical systems.

Troubleshooting and repairing electric motors is a valuable skill for electrical professionals. Here's a general guide to the process, including identifying common motor problems and performing motor maintenance tasks.

1. Identifying common motor problems:
   - No start or slow start: If the motor doesn't start or starts slowly, check for issues with the power supply, fuses, or overload protection devices. Additionally, consider inspecting the motor's windings and capacitors.
   - Overheating: An overheating motor can indicate issues with ventilation, excessive load, or bearing failure. Inspect the motor's cooling system and ensure it's not overloaded.
   - Excessive noise or vibration: Unusual noise or vibration may point to mechanical issues, such as misalignment, worn bearings, or unbalanced components.
   - Intermittent operation: If the motor operates inconsistently, there might be issues with loose connections, worn brushes, or faulty contactors.

2. Motor maintenance tasks:
   - Visual inspection: Regularly inspect the motor for signs of wear or damage, such as cracks, corrosion, or loose components.
   - Lubrication: Keep bearings properly lubricated to reduce friction and extend motor life. Follow the manufacturer's recommendations for lubrication intervals and types.
   - Cleanliness: Keep the motor clean and free of debris, particularly in cooling and ventilation areas, to ensure efficient heat dissipation.
   - Check brushes and commutators: Inspect the brushes and commutators of DC and universal motors for wear, and replace them as needed.

3. Troubleshooting steps:
   - Disconnect power: Before working on a motor, always disconnect and lock out the power supply to ensure safety.
   - Inspect the motor: Perform a visual inspection to identify any obvious signs of damage or wear.
   - Measure voltage and current: Use a multimeter to check the voltage and current at the motor terminals to ensure it's within the manufacturer's specifications.
   - Test motor windings: Use a multimeter or megohmmeter to check the motor windings for continuity and insulation resistance. Compare the readings to the manufacturer's specifications or NEMA guidelines.

- Check mechanical components: Inspect bearings, belts, and other mechanical components for signs of wear or damage.
4. Repairing the motor:
    - Replace damaged components: If any components are damaged or worn, replace them as needed. This may include bearings, brushes, or capacitors.
    - Realign the motor: If misalignment is causing excessive noise or vibration, realign the motor with the driven load.
    - Balance rotating components: If unbalanced components are causing vibration, balance them according to the manufacturer's guidelines.
    - Rewind the motor: If the motor windings are damaged, consider rewinding the motor or replacing it, depending on the extent of the damage and cost of repair.

By following these steps, you can effectively troubleshoot and repair electric motors, ensuring the reliable and efficient operation of your electrical systems.

Troubleshooting and maintaining lighting systems is essential for ensuring the proper functioning and efficiency of residential and commercial installations. Here are some methods for diagnosing and repairing common lighting issues, such as flickering lights, burnt-out lamps, and malfunctioning fixtures:

1. Flickering lights:
    - Check the connections: Loose connections can cause flickering. Inspect and tighten connections at the lamp, fixture, and electrical box.
    - Replace the lamp: Flickering may be due to an aging or defective lamp. Replace it with a new one to see if the issue resolves.
    - Inspect the fixture: A faulty fixture can also cause flickering. Examine the fixture for signs of wear or damage, and replace it if necessary.
    - Investigate dimmer switches: If the lights are connected to a dimmer switch, ensure it's compatible with the lamps being used. Replace the dimmer or lamp as needed.
2. Burnt-out lamps:
    - Replace the lamp: If a lamp burns out frequently, it could be due to a manufacturing defect or the wrong type of lamp for the fixture. Replace the lamp with a suitable one.
    - Check the voltage: A high voltage can cause lamps to burn out prematurely. Use a multimeter to check the voltage at the fixture and compare it to the lamp's rated voltage. If necessary, consult an electrician to address voltage issues.
    - Inspect the fixture: Examine the fixture for signs of overheating or other damage, and replace it if needed.
3. Malfunctioning fixtures:
    - Loose connections: Check and tighten all connections at the fixture, electrical box, and switch.

- Examine the wiring: Inspect the wiring for signs of wear or damage. Replace any damaged wires and ensure proper connections.
- Test the switch: Use a multimeter to test the switch for proper operation. Replace the switch if it's faulty.
- Inspect the ballast or driver: For fluorescent or LED lights, a malfunctioning fixture may be due to a faulty ballast or driver. Test these components and replace them if necessary.

4. General maintenance:
    - Regularly clean fixtures: Dust and dirt can accumulate on fixtures and reduce their efficiency. Clean them periodically to ensure optimal performance.
    - Inspect and replace lamps: Regularly check lamps for signs of wear or damage, and replace them as needed to maintain proper lighting levels.
    - Perform periodic inspections: Conduct routine visual inspections of the lighting system, looking for signs of wear or damage to fixtures, wiring, and other components.

By following these methods, you can effectively diagnose and repair common lighting issues, ensuring that your lighting systems remain functional and efficient.

# Practice Exam Section.

Welcome to the Practice Exam section of this study guide! The purpose of this section is to provide you with a collection of sample questions that simulate the format and content of a real exam. By working through these practice questions, you will gain a better understanding of the material, develop problem-solving skills, and build your confidence in tackling electrical installations and troubleshooting tasks.

We understand that learning is more effective when feedback is immediate, and that searching for answers in the back of the book can be frustrating and time-consuming. Therefore, we have chosen to include the correct answer and a detailed explanation right after each question. This way, you can quickly check your response, understand the reasoning behind the correct answer, and learn from any mistakes you might make.

This immediate feedback approach will help you reinforce your knowledge, clarify any misunderstandings, and identify areas where you may need to focus your studies further. By reviewing the explanations, you will gain a deeper understanding of the concepts and principles involved in electrical installations and troubleshooting, ensuring that you are well-prepared for your exam and real-world applications.

As you work through the practice exam questions, we encourage you to take your time and carefully read each question before attempting to answer. Think about the concepts and methods you have learned in the study guide, and try to apply them to each problem. Remember that practice makes perfect, so don't be discouraged if you make mistakes – learn from them and keep moving forward.

1. In a residential electrical system, the voltage across an electrical outlet is typically:
a) 12 V
b) 24 V
c) 120 V
d) 240 V


Answer: c) 120 V. Explanation: In most residential electrical systems, the standard voltage across an electrical outlet is 120 V. This voltage level is commonly used for powering household appliances and devices. Some larger appliances, such as electric dryers and stoves, may require 240 V.

2. Which of the following best describes the relationship between voltage, current, and resistance according to Ohm's Law?
a) Voltage = Current × Resistance
b) Current = Voltage × Resistance
c) Resistance = Voltage × Current
d) Current = Voltage / Resistance

Answer: a) Voltage = Current × Resistance
Explanation: Ohm's Law defines the relationship between voltage, current, and resistance in an electrical circuit. According to Ohm's Law, voltage (V) is equal to the product of current (I) and resistance (R), which can be expressed as V = I × R.

3. If the current in a circuit is 2 A and the resistance is 4 Ω, what is the power dissipated in the circuit?
a) 8 W
b) 16 W
c) 32 W
d) 64 W

Answer: b) 16 W
Explanation: To calculate the power (P) dissipated in a circuit, you can use the formula P = $I^2$ × R, where I is the current and R is the resistance. In this case, the current is 2 A and the resistance is 4 Ω, so the power dissipated is P = $(2 A)^2$ × 4 Ω = 16 W.

4. A device with a power rating of 1500 W is connected to a 120 V power source. What is the current drawn by the device?
a) 5 A
b) 10 A
c) 12.5 A
d) 15 A

Answer: c) 12.5 A
Explanation: To calculate the current (I) drawn by a device, you can use the formula P = V × I, where P is the power rating and V is the voltage. Rearranging the formula, we have I = P / V. In this case, the power rating is 1500 W and the voltage is 120 V, so the current drawn is I = 1500 W / 120 V = 12.5 A.

5. Which of the following materials typically has the highest electrical resistance?
a) Copper
b) Aluminum
c) Gold
d) Rubber

Answer: d) Rubber
Explanation: Among the given options, rubber has the highest electrical resistance. Copper, aluminum, and gold are all good conductors of electricity, which means they have low resistance. In contrast, rubber is an insulator, which means it has high resistance and does not allow electrical current to flow easily.

6. A technician needs to calculate the current flowing through a circuit with a voltage of 12 V and a resistance of 3 Ω. What is the current in the circuit according to Ohm's Law?
a) 3 A
b) 4 A
c) 9 A
d) 36 A

Answer: b) 4 A
Explanation: Ohm's Law states that voltage (V) equals current (I) times resistance (R). In this case, we need to find the current, so we can rearrange the equation as I = V / R. The voltage is 12 V, and the resistance is 3 Ω, so the current is I = 12 V / 3 Ω = 4 A.

7. An electric heater with a resistance of 10 Ω is connected to a 240 V power supply. What is the power consumed by the heater?
a) 5.76 kW
b) 2.4 kW
c) 576 W
d) 24 W

Answer: c) 576 W. Explanation: First, we need to find the current using Ohm's Law, which is I = V / R. In this case, the voltage is 240 V, and the resistance is 10 Ω, so the current is I = 240 V / 10 Ω = 24 A. To calculate the power consumed, we can use the formula P = V × I. Therefore, the power consumed is P = 240 V × 24 A = 576 W.

8. In an electrical circuit, the current flowing through a resistor is 6 A, and the voltage across the resistor is 18 V. What is the resistance of the resistor?
a) 1 Ω
b) 2 Ω
c) 3 Ω
d) 4 Ω

Answer: c) 3 Ω. Explanation: Ohm's Law states that voltage (V) equals current (I) times resistance (R). In this case, we need to find the resistance, so we can rearrange the equation as R = V / I. The voltage is 18 V, and the current is 6 A, so the resistance is R = 18 V / 6 A = 3 Ω.

9. A 120 V power source is connected to a device with a resistance of 20 Ω. If the resistance is doubled, how will the current change?
a) It will double.
b) It will halve.
c) It will remain the same.
d) It will quadruple.

Answer: b) It will halve.
Explanation: Using Ohm's Law, I = V / R. If the resistance is doubled, the new resistance is 40 Ω. The new current can be found as I = 120 V / 40 Ω = 3 A. Since the original current was 6 A (I = 120 V / 20 Ω), the current has halved when the resistance doubled.

10. A circuit has a current of 5 A and a resistance of 4 Ω. If the resistance is increased by 50%, what will be the new current in the circuit?
a) 2.5 A
b) 3.33 A
c) 7.5 A
d) 10 A


Answer: b) 3.33 A
Explanation: If the resistance is increased by 50%, the new resistance is 6 Ω (4 Ω × 1.5). Using Ohm's Law, we can find the new current as I = V / R. Since the voltage remains constant, the equation

11. According to Kirchhoff's Voltage Law (KVL), the sum of the voltage drops around a closed loop in a circuit is equal to:
a) Zero
b) The total voltage supplied
c) The total resistance
d) The total current


Answer: a) Zero
Explanation: Kirchhoff's Voltage Law states that the algebraic sum of the voltages around a closed loop in a circuit is equal to zero. This means that the sum of the voltage drops across the components in the loop is equal to the sum of the voltage sources in the loop.

12. When applying Kirchhoff's Current Law (KCL) to a node in an electrical circuit, which statement is true?
a) The sum of the currents entering the node equals the sum of the currents leaving the node.
b) The sum of the currents entering the node equals the total resistance at the node.
c) The sum of the currents leaving the node equals the total voltage at the node.
d) The sum of the currents entering the node equals the total power at the node.


Answer: a) The sum of the currents entering the node equals the sum of the currents leaving the node.
Explanation: Kirchhoff's Current Law states that the algebraic sum of the currents entering a node is equal to the sum of the currents leaving the node. This means that the total current flowing into the node must equal the total current flowing out of the node.

13. In a parallel circuit with three resistors, if one resistor is removed, what will happen to the total current in the circuit?
a) It will increase.
b) It will decrease.
c) It will remain the same.
d) It cannot be determined.

Answer: b) It will decrease.
Explanation: In a parallel circuit, the total current is the sum of the currents flowing through each resistor. If one resistor is removed, the current flowing through that resistor is no longer part of the total current, resulting in a decrease in the total current.

14. When applying Kirchhoff's Voltage Law to a series circuit with three resistors and a voltage source, the voltage across each resistor is determined by:
a) The resistance of each resistor.
b) The total resistance of the circuit.
c) The total current in the circuit.
d) The total power of the circuit.

Answer: c) The total current in the circuit.
Explanation: In a series circuit, the current is the same through all the resistors. According to Ohm's Law, the voltage across each resistor is the product of the current and the resistance ($V = I \times R$). Therefore, the voltage across each resistor depends on the total current in the circuit.

15. If the sum of currents at a junction in an electrical circuit is not equal to zero, which of the following is a possible explanation?
a) There is a measurement error.
b) Kirchhoff's Current Law does not apply.
c) The circuit is not properly grounded.
d) The circuit contains a transformer.

Answer: a) There is a measurement error.
Explanation: If the sum of currents at a junction does not equal zero, it could be due to a measurement error or inaccuracies in the values used in the calculation. Kirchhoff's Current Law should always hold true, as it is based on the conservation of charge.

16. In a series circuit with three identical resistors, the total resistance is:
a) Equal to the resistance of one resistor.
b) Half the resistance of one resistor.
c) Three times the resistance of one resistor.
d) Nine times the resistance of one resistor.

Answer: c) Three times the resistance of one resistor.
Explanation: In a series circuit, the total resistance is equal to the sum of the individual resistances. Since all three resistors have the same resistance, the total resistance is three times the resistance of one resistor.

17. What happens to the total resistance in a parallel circuit if a new resistor is added?
a) It increases.
b) It decreases.
c) It remains the same.
d) It cannot be determined.

Answer: b) It decreases.
Explanation: In a parallel circuit, the total resistance can be calculated using the formula $1/R_t = 1/R_1 + 1/R_2 + ... + 1/R_n$. Adding a new resistor to the parallel circuit results in a decrease in the total resistance, as more pathways are provided for the current to flow.

18. In a parallel circuit with multiple resistors, what is true about the voltage across each resistor?
a) The voltage is the same across all resistors.
b) The voltage is different across each resistor.
c) The voltage is proportional to the resistance.
d) The voltage is inversely proportional to the resistance.

Answer: a) The voltage is the same across all resistors.
Explanation: In a parallel circuit, the voltage across each resistor is equal to the voltage supplied by the source. This is because each resistor is connected directly to the voltage source, and they all share the same potential difference.

19. If a light bulb in a series circuit with multiple light bulbs burns out, what will happen to the other light bulbs?
a) They will become brighter.
b) They will become dimmer.
c) They will turn off.
d) Their brightness will remain the same.

Answer: c) They will turn off.
Explanation: In a series circuit, the current flows through all components in a single path. If one light bulb burns out, it breaks the circuit, and the current cannot flow through the other light bulbs, causing them to turn off.

20. What is the primary advantage of a parallel circuit compared to a series circuit in practical applications?
a) Parallel circuits have higher total resistance.
b) Parallel circuits have lower total resistance.
c) Parallel circuits can maintain the same voltage across each component.
d) Parallel circuits can maintain the same current through each component.

Answer: c) Parallel circuits can maintain the same voltage across each component.
Explanation: In a parallel circuit, the voltage across each component is the same, as they are all directly connected to the voltage source. This is advantageous in practical applications, as it ensures that each component receives the required voltage regardless of the number of components in the circuit.

21. Which of the following best describes the primary difference between alternating current (AC) and direct current (DC)?
a) AC flows in a single direction, while DC changes direction periodically.
b) AC changes direction periodically, while DC flows in a single direction.
c) AC has a constant amplitude, while DC has a variable amplitude.
d) AC has a variable amplitude, while DC has a constant amplitude.

Answer: b) AC changes direction periodically, while DC flows in a single direction.
Explanation: Alternating current (AC) changes direction periodically, while direct current (DC) flows in a single, constant direction. This is the primary difference between AC and DC.

22. What is the primary advantage of alternating current (AC) over direct current (DC) for power transmission over long distances?
a) AC has lower power loss during transmission.
b) AC has higher power loss during transmission.
c) AC can be easily converted to different voltages.
d) AC has a constant amplitude.

Answer: a) AC has lower power loss during transmission.
Explanation: The primary advantage of alternating current (AC) over direct current (DC) for power transmission over long distances is its lower power loss. AC can be transmitted at higher voltages, which reduces the current and therefore the power loss in the transmission lines due to resistance.

23. The number of cycles of an AC waveform that occur in one second is known as:
a) Amplitude.
b) Frequency.
c) Phase.
d) Period.

Answer: b) Frequency.
Explanation: The number of cycles of an AC waveform that occur in one second is known as frequency, which is measured in hertz (Hz).

24. Which of the following properties of an AC waveform represents the maximum value of the voltage or current?
a) Amplitude.
b) Frequency.
c) Phase.
d) Period.

Answer: a) Amplitude.
Explanation: Amplitude is the property of an AC waveform that represents the maximum value of the voltage or current. It indicates the peak value of the voltage or current in the waveform.

25. In a three-phase AC system, the phase difference between any two adjacent waveforms is:
a) 30 degrees.
b) 60 degrees.
c) 90 degrees.
d) 120 degrees.

Answer: d) 120 degrees.
Explanation: In a three-phase AC system, the phase difference between any two adjacent waveforms is 120 degrees. This ensures that the power delivery remains constant and reduces the need for large neutral conductors.

26. Capacitors and inductors are both energy storage devices in electrical circuits. Which of the following statements best describes their primary functions?
a) Capacitors store energy in electric fields, while inductors store energy in magnetic fields.
b) Capacitors store energy in magnetic fields, while inductors store energy in electric fields.
c) Both capacitors and inductors store energy in electric fields.
d) Both capacitors and inductors store energy in magnetic fields.

Answer: a) Capacitors store energy in electric fields, while inductors store energy in magnetic fields.
Explanation: Capacitors store energy in electric fields between their plates, while inductors store energy in magnetic fields generated by the current flowing through their coils. This is the primary difference in their energy storage mechanisms.

27. In an AC circuit, how does the impedance of a capacitor change as the frequency increases?
a) Impedance increases.
b) Impedance decreases.
c) Impedance remains constant.
d) Impedance becomes zero.


Answer: b) Impedance decreases.
Explanation: In an AC circuit, the impedance of a capacitor decreases as the frequency increases. This is because the capacitive reactance (Xc) is inversely proportional to the frequency (Xc = 1/(2πfC), where f is frequency and C is capacitance).

28. In an AC circuit, how does the impedance of an inductor change as the frequency increases?
a) Impedance increases.
b) Impedance decreases.
c) Impedance remains constant.
d) Impedance becomes zero.


Answer: a) Impedance increases.
Explanation: In an AC circuit, the impedance of an inductor increases as the frequency increases. This is because the inductive reactance (XL) is directly proportional to the frequency (XL = 2πfL, where f is frequency and L is inductance).

29 In an AC circuit containing both a capacitor and an inductor, which of the following conditions results in resonance?
a) Capacitive reactance equals inductive reactance.
b) Capacitive reactance is greater than inductive reactance.
c) Capacitive reactance is less than inductive reactance.
d) Capacitive reactance and inductive reactance are both zero.


Answer: a) Capacitive reactance equals inductive reactance.
Explanation: In an AC circuit containing both a capacitor and an inductor, resonance occurs when capacitive reactance equals inductive reactance. At resonance, the impedance of the circuit is minimized, and the current reaches its maximum value.

30. In a purely capacitive AC circuit, the voltage and current are:
a) In phase.
b) 90 degrees out of phase, with voltage leading current.
c) 90 degrees out of phase, with current leading voltage.
d) 180 degrees out of phase.

Answer: c) 90 degrees out of phase, with current leading voltage.
Explanation: In a purely capacitive AC circuit, the voltage and current are 90 degrees out of phase, with current leading voltage. This is because the current in a capacitive circuit changes more rapidly than the voltage across the capacitor.

31. Impedance in AC circuits is a measure of:
a) The rate of change of current with respect to voltage.
b) The opposition to current flow.
c) The energy stored in electric and magnetic fields.
d) The amount of charge stored per unit voltage.

Answer: b) The opposition to current flow.
Explanation: Impedance is the opposition to the flow of alternating current in an AC circuit. It takes into account the effects of resistance, inductive reactance, and capacitive reactance on the current flow.

32. In a series AC circuit containing resistors, capacitors, and inductors, the total impedance (Z) is calculated as:
a) $Z = R + X_c + X_L$
b) $Z = R - (X_c - X_L)$
c) $Z = R - (X_L - X_c)$
d) $Z = \sqrt{(R^2) + (X_L - X_c)^2}$

Answer: d) $Z = \sqrt{(R^2) + (X_L - X_c)^2}$
Explanation: In a series AC circuit, the total impedance is the square root of the sum of the resistance squared and the difference between inductive reactance and capacitive reactance squared ($Z = \sqrt{(R^2) + (X_L - X_c)^2}$).

33. In a parallel AC circuit containing only resistors and capacitors, the total impedance (Z) can be calculated using:
a) $Z = \sqrt{(1/R^2) + (1/X_c^2)}$
b) $Z = 1 / (1/R + 1/X_c)$
c) $Z = R + X_c$
d) $Z = R * X_c / \sqrt{(R^2) + (X_c^2)}$

Answer: b) $Z = 1 / (1/R + 1/X_c)$
Explanation: In a parallel AC circuit containing only resistors and capacitors, the total impedance can be found using the formula $Z = 1 / (1/R + 1/X_c)$.

34. In an AC circuit, which of the following has the least effect on impedance at high frequencies?
a) Resistors
b) Capacitors
c) Inductors
d) Transformers

Answer: a) Resistors
Explanation: At high frequencies, the impedance of capacitors decreases while the impedance of inductors increases. Resistors, however, do not change their impedance based on frequency, making their effect on impedance the least dependent on frequency.

35. When comparing two inductors in an AC circuit, the one with a higher inductance value will have:
a) Higher impedance at the same frequency.
b) Lower impedance at the same frequency.
c) Equal impedance at the same frequency.
d) Impedance that is independent of frequency.

Answer: a) Higher impedance at the same frequency.
Explanation: The inductive reactance (XL) is directly proportional to the inductance (L) and frequency (f) of the AC circuit ($XL = 2\pi fL$). Therefore, at the same frequency, an inductor with a higher inductance value will have a higher impedance.

36. Power factor in an AC circuit is defined as the:
a) Ratio of real power to apparent power.
b) Ratio of apparent power to real power.
c) Ratio of reactive power to apparent power.
d) Ratio of apparent power to reactive power.

Answer: a) Ratio of real power to apparent power.
Explanation: The power factor is the ratio of real power (P) to apparent power (S) in an AC circuit. It is a measure of how effectively the electrical power is being converted into useful work.

37. A power factor of 1.0 indicates that:
a) All the power is consumed as reactive power.
b) All the power is consumed as real power.
c) No power is being consumed.
d) The circuit has only resistive elements.

Answer: b) All the power is consumed as real power.
Explanation: A power factor of 1.0 means that all the power drawn from the source is being consumed as real power, which is used for useful work. This situation occurs in purely resistive circuits.

38. A lagging power factor in an AC circuit is usually associated with:
a) Resistive loads.
b) Capacitive loads.
c) Inductive loads.
d) Nonlinear loads.

Answer: c) Inductive loads.
Explanation: A lagging power factor is typically associated with inductive loads, such as motors and transformers, where the current lags behind the voltage due to the presence of inductance in the circuit.

39. Power factor correction is typically accomplished by:
a) Increasing the resistance in the circuit.
b) Decreasing the resistance in the circuit.
c) Adding capacitors or inductors to the circuit.
d) Removing capacitors or inductors from the circuit.

Answer: c) Adding capacitors or inductors to the circuit.
Explanation: Power factor correction is usually achieved by adding capacitors or inductors to the circuit. This compensates for the reactive power and brings the power factor closer to 1, which results in increased energy efficiency.

40. The primary benefit of improving power factor in an AC circuit is:
a) Reduced energy consumption.
b) Increased power quality.
c) Reduced power losses in transmission lines.
d) Decreased equipment maintenance costs.

Answer: c) Reduced power losses in transmission lines.
Explanation: Improving the power factor in an AC circuit primarily reduces power losses in transmission lines due to reduced reactive power. This leads to increased energy efficiency and may also result in reduced energy consumption and lower utility costs.

41. The primary purpose of a transformer is to:
a) Convert AC to DC.
b) Convert DC to AC.
c) Change voltage levels in AC circuits.
d) Measure electrical power consumption.

Answer: c) Change voltage levels in AC circuits.
Explanation: Transformers are used to change voltage levels in AC circuits. They can step up or step down voltages, depending on the application.

42. In a step-up transformer:
a) The primary coil has more turns than the secondary coil.
b) The primary coil has fewer turns than the secondary coil.
c) The primary coil and secondary coil have the same number of turns.
d) The number of turns in the primary coil is unrelated to the secondary coil.

Answer: b) The primary coil has fewer turns than the secondary coil.
Explanation: In a step-up transformer, the secondary coil has more turns than the primary coil, resulting in a higher voltage on the secondary side.

43. Transformers are used in impedance matching to:
a) Increase the load impedance.
b) Decrease the load impedance.
c) Maximize power transfer between circuits.
d) Minimize power transfer between circuits.

Answer: c) Maximize power transfer between circuits.
Explanation: Transformers are used in impedance matching to maximize power transfer between circuits by ensuring that the load impedance is equal to the source impedance.

44. Which of the following transformer types provides electrical isolation between its primary and secondary windings?
a) Auto-transformer.
b) Isolation transformer.
c) Step-up transformer.
d) Step-down transformer.

Answer: b) Isolation transformer.
Explanation: Isolation transformers provide electrical isolation between their primary and secondary windings. This can be useful in protecting sensitive equipment from voltage spikes and noise.

45. The efficiency of a transformer is primarily affected by:
a) Resistive losses in the windings.
b) Capacitive losses in the windings.
c) Reactive power losses in the core.
d) Both resistive losses in the windings and reactive power losses in the core.

Answer: d) Both resistive losses in the windings and reactive power losses in the core.
Explanation: The efficiency of a transformer is primarily affected by resistive losses in the windings, which are due to the resistance of the wire, and reactive power losses in the core, which are due to hysteresis and eddy currents.

46. The main advantage of a three-phase electrical system compared to a single-phase system is:
a) Easier conversion to DC.
b) Lower power consumption.
c) Greater power transmission capacity.
d) Simplified wiring.

Answer: c) Greater power transmission capacity.
Explanation: Three-phase electrical systems provide greater power transmission capacity compared to single-phase systems, making them more efficient for power distribution and industrial applications.

47. In a balanced three-phase system, the sum of the three instantaneous phase voltages is:
a) Equal to zero.
b) Equal to the peak voltage.
c) Equal to the phase voltage.
d) Indeterminate.

Answer: a) Equal to zero.
Explanation: In a balanced three-phase system, the sum of the three instantaneous phase voltages is zero. This is because the phase voltages are equal in magnitude but have a 120° phase difference.

48. What is the phase difference between the line voltages in a balanced three-phase system?
a) 60°
b) 120°
c) 180°
d) 240°

Answer: b) 120°
Explanation: In a balanced three-phase system, the phase difference between the line voltages is 120°. This ensures a constant power transfer in the system.

49. A three-phase motor is more efficient than a single-phase motor because:
a) It has fewer windings.
b) It requires less maintenance.
c) It produces a smoother torque output.
d) It has a higher power factor.

Answer: c) It produces a smoother torque output.
Explanation: Three-phase motors are more efficient than single-phase motors because they produce a smoother torque output, reducing vibrations and improving overall performance.

50. A wye-connected three-phase system is characterized by:
a) A common neutral point.
b) No neutral point.
c) Equal voltages between all phases.
d) Higher line currents than delta-connected systems.

Answer: a) A common neutral point.
Explanation: In a wye-connected three-phase system, the windings are connected in such a way that they form a common neutral point. This configuration allows for both line-to-line and line-to-neutral voltage connections.

51. Schematic diagrams are primarily used to:
a) Show the physical layout of a circuit.
b) Represent the logical structure of a circuit.
c) Display the power distribution in a circuit.
d) Illustrate the wiring connections in a circuit.

Answer: b) Represent the logical structure of a circuit.
Explanation: Schematic diagrams use standardized symbols to represent the logical structure of a circuit, making it easier to understand the relationships between components.

52. Wiring diagrams differ from schematic diagrams in that they:
a) Use different symbols for components.
b) Focus on the physical layout of a circuit.
c) Only represent single-phase circuits.
d) Are used exclusively for power distribution.

Answer: b) Focus on the physical layout of a circuit.
Explanation: Wiring diagrams focus on the physical layout of a circuit, showing the actual connections between components and how they are wired together.

53. Which type of electrical diagram is best suited for illustrating the overall organization of a complex system?
a) Schematic diagram
b) Wiring diagram
c) Block diagram
d) Circuit diagram

Answer: c) Block diagram.
Explanation: Block diagrams are used to illustrate the overall organization of a complex system, showing the functional relationships between major components or subsystems.

54. In a schematic diagram, a resistor is typically represented by:
a) A zig-zag line.
b) A straight line.
c) A circle.
d) A rectangle.

Answer: a) A zig-zag line.
Explanation: In schematic diagrams, resistors are usually represented by a zig-zag line, making it easy to identify them among other components.

55. What is the main purpose of using standardized symbols in electrical diagrams?
a) To make the diagrams more visually appealing.
b) To simplify the design process for engineers.
c) To ensure consistent interpretation and understanding.
d) To save time in drawing complex circuits.

Answer: c) To ensure consistent interpretation and understanding.
Explanation: The main purpose of using standardized symbols in electrical diagrams is to ensure consistent interpretation and understanding of the diagrams, allowing professionals to communicate and collaborate effectively.

56. The primary purpose of grounding in electrical systems is to:
a) Reduce energy consumption.
b) Protect equipment from power surges.
c) Provide a safe path for fault current to flow.
d) Improve the efficiency of electrical devices.

Answer: c) Provide a safe path for fault current to flow.
Explanation: Grounding provides a safe path for fault current to flow, helping to prevent electric shock and minimize the risk of fire in the event of a short circuit or other fault condition.

57. A short circuit occurs when:
a) The current in a circuit exceeds the rated capacity.
b) An unintended connection is made between two points of different voltage.
c) Resistance in a circuit becomes too high.
d) A circuit is left open, interrupting the flow of current.

Answer: b) An unintended connection is made between two points of different voltage.
Explanation: A short circuit occurs when an unintended connection is made between two points of different voltage, causing a large current flow that can lead to overheating, fire, or damage to equipment.

58. Which of the following devices is designed to protect a circuit by opening it when the current exceeds a predetermined limit?
a) Transformer
b) Fuse
c) Inductor
d) Capacitor

Answer: b) Fuse
Explanation: Fuses are designed to protect a circuit by opening it when the current exceeds a predetermined limit, effectively cutting off the flow of current to prevent damage to equipment and reduce the risk of fire.

59. The primary difference between a fuse and a circuit breaker is that:
a) Fuses are used only in residential applications, while circuit breakers are used in commercial and industrial settings.
b) Fuses provide a higher level of protection than circuit breakers.
c) Circuit breakers can be reset after tripping, while fuses must be replaced.
d) Circuit breakers are faster-acting than fuses.

Answer: c) Circuit breakers can be reset after tripping, while fuses must be replaced.
Explanation: The primary difference between a fuse and a circuit breaker is that circuit breakers can be reset after tripping, while fuses must be replaced once they have blown.

60. Ground fault circuit interrupters (GFCIs) are designed to protect against:
a) Overloads.
b) Short circuits.
c) Ground faults.
d) Voltage spikes.

Answer: c) Ground faults.
Explanation: Ground fault circuit interrupters (GFCIs) are designed to protect against ground faults, which occur when current leaks from a circuit and flows to the ground, potentially causing electric shock. GFCIs monitor the flow of current and quickly disconnect the power if a ground fault is detected.

61. The National Electrical Code (NEC) was first published in:
a) 1873
b) 1897
c) 1920
d) 1941

Answer: b) 1897
Explanation: The NEC was first published in 1897 as a means to ensure electrical safety and uniformity in electrical installations across the United States.

62. The primary purpose of the National Electrical Code (NEC) is to:
a) Specify the materials to be used in all electrical installations.
b) Establish minimum requirements for safe electrical design, installation, and inspection.
c) Set pricing standards for electrical components and equipment.
d) Provide guidelines for energy efficiency in electrical systems.

Answer: b) Establish minimum requirements for safe electrical design, installation, and inspection.
Explanation: The NEC's primary purpose is to establish minimum requirements for safe electrical design, installation, and inspection to protect people and property from electrical hazards.

63. The NEC is updated and published by which organization?
a) Institute of Electrical and Electronics Engineers (IEEE)
b) Occupational Safety and Health Administration (OSHA)
c) National Fire Protection Association (NFPA)
d) American National Standards Institute (ANSI)

Answer: c) National Fire Protection Association (NFPA)
Explanation: The NEC is updated and published by the National Fire Protection Association (NFPA), a nonprofit organization dedicated to reducing the burden of fire and other hazards.

64. How often is the National Electrical Code (NEC) typically updated?
a) Every year
b) Every two years
c) Every three years
d) Every five years

Answer: c) Every three years
Explanation: The NEC is typically updated every three years to reflect advancements in technology, safety research, and the evolving needs of the electrical industry.

65. Which of the following statements about the National Electrical Code (NEC) is true?
a) The NEC is a federal law that applies to all states.
b) The NEC is a voluntary standard that has no legal significance.
c) Adoption of the NEC is determined at the state or local level.
d) The NEC is enforced by the federal government.

Answer: c) Adoption of the NEC is determined at the state or local level.
Explanation: Although the NEC is a widely recognized safety standard, it is not a federal law. Adoption of the NEC is determined at the state or local level, and its provisions are enforced by local authorities.

66. The National Electrical Code (NEC) is divided into:
a) Sections
b) Articles
c) Chapters
d) All of the above

Answer: d) All of the above
Explanation: The NEC is divided into chapters, which are further divided into articles. Articles are then broken down into sections and subsections for a logical organization of the content.

67. How many chapters does the National Electrical Code (NEC) typically contain?
a) 9
b) 12
c) 15
d) 18

Answer: a) 9
Explanation: The NEC is typically organized into 9 chapters, with each chapter focusing on a specific aspect of electrical design, installation, or inspection.

68. Which chapter of the NEC is dedicated to wiring methods and materials?
a) Chapter 1
b) Chapter 2
c) Chapter 3
d) Chapter 4
Answer: c) Chapter 3
Explanation: Chapter 3 of the NEC is dedicated to wiring methods and materials, providing guidelines and requirements for various types of wiring systems and components.

69. The NEC's Article 100 contains:
a) Definitions
b) General requirements for electrical installations
c) Wiring methods and materials
d) Equipment for general use


Answer: a) Definitions
Explanation: Article 100 of the NEC contains definitions for terms used throughout the code, helping to ensure a clear and consistent understanding of the requirements.

70. Which of the following NEC articles focuses on grounding and bonding?
a) Article 240
b) Article 250
c) Article 300
d) Article 310


Answer: b) Article 250
Explanation: Article 250 of the NEC focuses on grounding and bonding, providing requirements for proper grounding and bonding of electrical systems to ensure safety and minimize potential hazards.

71. A grounded conductor is:
a) A conductor connected to the ground
b) A conductor carrying current under normal operating conditions
c) A conductor that is insulated from the ground
d) A conductor carrying current only during a fault condition

Answer: b) A conductor carrying current under normal operating conditions
Explanation: A grounded conductor, often referred to as the neutral conductor, carries current under normal operating conditions and is connected to the grounding system at the service entrance or source.

72. The purpose of a grounding electrode is to:
a) Carry current under normal operating conditions
b) Provide a connection between the electrical system and the earth
c) Protect electrical devices from overcurrent conditions
d) Distribute power to individual branch circuits

Answer: b) Provide a connection between the electrical system and the earth
Explanation: A grounding electrode serves as a connection between the electrical system and the earth, establishing a ground reference for the system and helping to dissipate fault currents safely.

73. A branch circuit is defined as:
a) A circuit that connects the service entrance to a main distribution panel
b) A circuit that distributes power from a transformer to multiple load centers
c) A circuit that connects a power source to a single load or group of loads
d) A circuit that connects multiple devices in parallel

Answer: c) A circuit that connects a power source to a single load or group of loads
Explanation: A branch circuit is a circuit that connects a power source (such as a panelboard) to a single load or group of loads, providing power to devices or equipment.

74. A feeder is responsible for:
a) Connecting a power source to a single load or group of loads
b) Distributing power from a main distribution panel to branch circuits
c) Connecting a power source to a main distribution panel
d) Connecting devices in series within a branch circuit

Answer: b) Distributing power from a main distribution panel to branch circuits
Explanation: A feeder is an electrical conductor that distributes power from a main distribution panel or other source to branch circuits or devices, providing power to multiple loads within a system.

75. The term "grounding" refers to:
a) The process of connecting a conductor to the earth
b) The process of connecting a conductor to a grounded object
c) The process of connecting a conductor to the power source
d) The process of connecting a conductor to the grounding electrode

Answer: a) The process of connecting a conductor to the earth
Explanation: Grounding refers to the process of establishing a connection between a conductor or an electrical system and the earth, providing a reference point for voltage levels and a path for fault currents to dissipate safely.

76. According to the NEC, the purpose of bonding is to:
a) Carry fault current back to the source
b) Establish a connection to the earth
c) Protect equipment from overcurrent conditions
d) Ensure all metal parts of an electrical system are at the same potential

Answer: d) Ensure all metal parts of an electrical system are at the same potential
Explanation: Bonding ensures that all metal parts of an electrical system are at the same potential, helping to prevent voltage differences that could cause electric shock or damage to equipment.

77. The primary purpose of overcurrent protection devices, such as fuses and circuit breakers, is to:
a) Regulate voltage levels
b) Protect conductors from excessive current
c) Provide a path to ground for fault currents
d) Limit the number of devices on a circuit

Answer: b) Protect conductors from excessive current
Explanation: Overcurrent protection devices are designed to protect conductors from excessive current that could cause overheating, insulation damage, and potential fire hazards.

78. When sizing conductors, the NEC requires that the minimum conductor size be based on:
a) The maximum continuous load
b) The maximum fault current
c) 125% of the continuous load plus the non-continuous load
d) The sum of the continuous and non-continuous loads

Answer: c) 125% of the continuous load plus the non-continuous load
Explanation: According to the NEC, the minimum conductor size should be based on 125% of the continuous load plus the non-continuous load to ensure that the conductor can safely carry the expected current without overheating or damage.

79. Grounding electrode conductors must be sized to:
a) Carry the maximum fault current
b) Provide a low impedance path to ground
c) Ensure proper operation of overcurrent protection devices
d) Withstand the maximum fault current without damage

Answer: d) Withstand the maximum fault current without damage
Explanation: Grounding electrode conductors must be sized to withstand the maximum fault current without damage, ensuring a safe and effective path for fault currents to dissipate.

80. To ensure proper grounding, the NEC requires that a grounding electrode system have a resistance to ground of:
a) 25 ohms or less
b) 50 ohms or less
c) 100 ohms or less
d) There is no specific resistance requirement

Answer: a) 25 ohms or less
Explanation: The NEC requires that a grounding electrode system have a resistance to ground of 25 ohms or less to ensure a low impedance path for fault currents and proper grounding of the electrical system.

81. According to the NEC, when installing electrical metallic tubing (EMT) in a dry location, which type of conductor insulation is suitable?
a) TW
b) THHN
c) UF
d) NM

Answer: b) THHN
Explanation: THHN (Thermoplastic High Heat-resistant Nylon-coated) insulation is suitable for use in dry locations, such as with EMT installations. It is resistant to heat and moisture, making it appropriate for this application.

82. The NEC requires that junction boxes be accessible:
a) Only during initial installation
b) Only during maintenance
c) At all times, without removing any building parts
d) Only when disconnecting the circuit

Answer: c) At all times, without removing any building parts
Explanation: According to the NEC, junction boxes must be accessible at all times without removing any building parts, ensuring that they can be easily accessed for maintenance, inspections, or modifications.

83. Which type of cable is suitable for use in damp or wet locations, according to the NEC?
a) Nonmetallic sheathed cable (NM)
b) Underground feeder (UF)
c) Thermoplastic high heat-resistant (THHN)
d) Service entrance (SE)

Answer: b) Underground feeder (UF)
Explanation: Underground feeder (UF) cable is suitable for use in damp or wet locations, as it is designed with a moisture-resistant insulation that can withstand exposure to water and damp conditions.

84. According to the NEC, when installing a raceway, the maximum number of bends between pull points should not exceed:
a) 180 degrees
b) 270 degrees
c) 360 degrees
d) 450 degrees

Answer: c) 360 degrees
Explanation: The NEC specifies that the maximum number of bends between pull points in a raceway should not exceed 360 degrees. This requirement helps ensure that cables can be easily pulled through the raceway without excessive tension or damage.

85. For wiring methods in hazardous locations, the NEC requires the use of:
a) Nonmetallic sheathed cable (NM)
b) Thermoplastic high heat-resistant (THHN)
c) Explosion-proof fittings and enclosures
d) Armored cable (AC)

Answer: c) Explosion-proof fittings and enclosures
Explanation: In hazardous locations, the NEC requires the use of explosion-proof fittings and enclosures to help prevent the ignition of flammable gases, vapors, or dust that may be present. This helps ensure the safety of electrical installations in these environments.

86. According to the NEC, ground-fault circuit interrupter (GFCI) protection is required for receptacles installed:
a) In all indoor locations
b) In commercial kitchens
c) Outdoors
d) Only in wet or damp locations

Answer: c) Outdoors
Explanation: The NEC requires GFCI protection for outdoor receptacles to protect against ground faults that can pose a serious safety risk, especially in wet or damp conditions where the risk of electrical shock is increased.

87. The NEC requires that a disconnecting means for a motor be located:
a) Within sight of the motor
b) In the same room as the motor
c) Within 25 feet of the motor
d) At the nearest distribution panel

Answer: a) Within sight of the motor
Explanation: According to the NEC, a disconnecting means for a motor must be located within sight of the motor. This ensures that personnel can easily and quickly disconnect power in case of an emergency or maintenance requirement.

88. When installing lighting fixtures in a clothes closet, the NEC requires a minimum clearance of _____ inches between the fixture and the nearest point of storage space.
a) 6
b) 12
c) 18
d) 24

Answer: b) 12
Explanation: The NEC mandates a minimum clearance of 12 inches between lighting fixtures and the nearest point of storage space in a clothes closet. This helps prevent the risk of fire caused by contact between the fixture and flammable materials.

89. According to the NEC, which type of switch is suitable for use with a fluorescent lighting circuit that operates at 277V?
a) A single-pole switch rated for 120V
b) A double-pole switch rated for 240V
c) A single-pole switch rated for 277V
d) A three-way switch rated for 120V

Answer: c) A single-pole switch rated for 277V
Explanation: The NEC requires a single-pole switch rated for 277V for use with a fluorescent lighting circuit operating at 277V. Using a switch with an appropriate voltage rating ensures safe operation and minimizes the risk of electrical hazards.

90. In a dwelling unit, the NEC requires that receptacle outlets be installed no more than _____ feet apart along a wall in habitable rooms.
a) 6
b) 8
c) 10
d) 12

Answer: d) 12
Explanation: According to the NEC, receptacle outlets in dwelling units must be installed no more than 12 feet apart along a wall in habitable rooms. This requirement helps ensure that an adequate number of receptacles are available, minimizing the need for extension cords and reducing the risk of electrical hazards.

91. In a hazardous location, the NEC requires electrical equipment to be:
a) Installed in explosion-proof enclosures
b) Rated for use in wet locations
c) Protected by ground-fault circuit interrupters
d) Installed with a minimum clearance of 18 inches from the floor

Answer: a) Installed in explosion-proof enclosures
Explanation: The NEC requires electrical equipment in hazardous locations to be installed in explosion-proof enclosures to minimize the risk of ignition and explosion caused by electrical sparks or heat generated by the equipment.

92. In healthcare facilities, the NEC requires that all branch circuits supplying patient care areas be provided with:
a) Ground-fault circuit interrupter protection
b) Arc-fault circuit interrupter protection
c) Isolated grounding
d) Surge protection devices

Answer: c) Isolated grounding
Explanation: The NEC requires isolated grounding in healthcare facilities to minimize electrical noise and interference that can affect sensitive medical equipment and ensure patient safety.

93. According to the NEC, the maximum voltage for a solar photovoltaic system in a residential application is:
a) 120V
b) 240V
c) 600V
d) 1000V

Answer: c) 600V
Explanation: The NEC specifies a maximum voltage of 600V for residential solar photovoltaic systems. This is to ensure the safety of occupants and electrical personnel who may work on or near the system.

94. Emergency systems, as defined by the NEC, are intended to provide power for:
a) Standby power generation
b) Fire alarm systems
c) Life safety and critical loads
d) Uninterruptible power supplies

Answer: c) Life safety and critical loads
Explanation: Emergency systems are defined by the NEC as systems that provide power for life safety and critical loads in the event of a utility power outage. These systems ensure that essential equipment continues to operate during an emergency.

95. When installing a battery system for an emergency power supply, the NEC requires that the battery room be:
a) Ventilated to prevent the buildup of hydrogen gas
b) Equipped with ground-fault circuit interrupter protection
c) Protected by a dedicated circuit breaker
d) Located within 25 feet of the main electrical panel

Answer: a) Ventilated to prevent the buildup of hydrogen gas
Explanation: The NEC requires battery rooms for emergency power supplies to be ventilated to prevent the buildup of hydrogen gas, which can be generated by lead-acid batteries during charging and can pose an explosion risk. Proper ventilation ensures that hydrogen gas is safely dispersed and minimizes the risk of an explosion.

96. According to the NEC, communication cables should not be installed in the same raceway, conduit, or enclosure with:
a) Class 2 circuits
b) Power-limited fire alarm circuits
c) Power conductors
d) Grounding conductors

Answer: c) Power conductors
Explanation: The NEC specifies that communication cables should not be installed in the same raceway, conduit, or enclosure with power conductors to avoid potential interference and safety issues caused by induced voltages or close proximity between the cables.

97. In a residential installation, the NEC requires a minimum of how many communication outlets per dwelling unit?
a) One
b) Two
c) Three
d) Four

Answer: a) One
Explanation: The NEC requires a minimum of one communication outlet per dwelling unit to ensure that occupants have access to telephone, television, or other communication services.

98. According to the NEC, the maximum length of a communications bonding conductor is:
a) 6 feet
b) 12 feet
c) 20 feet
d) No maximum length specified

Answer: b) 12 feet
Explanation: The NEC specifies a maximum length of 12 feet for a communications bonding conductor. This is to ensure that the communications grounding system is effective in dissipating any induced voltages, surges, or noise that may affect the performance of communication systems.

99. When installing communication wires in a suspended ceiling, the NEC requires the wires to be:
a) Supported by independent support wires
b) Attached to the ceiling grid
c) Run through metal conduit
d) Bundled together with cable ties

Answer: a) Supported by independent support wires
Explanation: The NEC requires communication wires in a suspended ceiling to be supported by independent support wires to ensure that the communication wires are secure and do not place undue stress on the ceiling grid.

100. The NEC requires a minimum separation distance between communication cables and lightning protection conductors of:
a) 2 inches
b) 6 inches
c) 12 inches
d) 24 inches

Answer: b) 6 inches
Explanation: The NEC requires a minimum separation distance of 6 inches between communication cables and lightning protection conductors. This is to minimize the risk of damage to the communication cables in the event of a lightning strike, which could cause induced voltages or arcing between the conductors.

101. The National Electrical Code (NEC) is updated on a regular basis to:
a) Increase the complexity of electrical installations
b) Address new technology and industry advancements
c) Generate revenue for the code developers
d) Make it difficult for electricians to pass exams

Answer: b) Address new technology and industry advancements
Explanation: The NEC is updated regularly to ensure that it remains relevant and effective in addressing new technology and industry advancements, ensuring the continued safety of electrical installations.

102. The typical revision cycle for the NEC is:
a) Every year
b) Every two years
c) Every three years
d) Every five years

Answer: c) Every three years
Explanation: The NEC is typically updated on a three-year revision cycle, ensuring that it stays current with advancements in technology and industry practices while providing a reasonable timeframe for adoption and implementation.

103. When a new edition of the NEC is released, it is automatically adopted by all jurisdictions in the United States.
a) True
b) False

Answer: b) False
Explanation: Local jurisdictions are responsible for adopting and enforcing the NEC, and they may choose to adopt a new edition, continue using an older edition, or adopt a modified version of the code based on their specific needs and requirements.

104. In order for a proposed change to be incorporated into the NEC, it must:
a) Be approved by a simple majority vote of the code-making panel
b) Be submitted by a licensed electrician
c) Pass through a rigorous public review process
d) Be endorsed by a major electrical manufacturer

Answer: c) Pass through a rigorous public review process
Explanation: Proposed changes to the NEC must go through a rigorous public review process, which includes submission of public inputs, public comment, and approval by the code-making panel. This process ensures that any changes to the code are carefully considered and based on input from a wide range of stakeholders.

105. The primary responsibility for enforcing the NEC within a local jurisdiction lies with:
a) The electrical contractor
b) The local building department or code enforcement agency
c) The utility company
d) The homeowner

Answer: b) The local building department or code enforcement agency
Explanation: The primary responsibility for enforcing the NEC within a local jurisdiction typically lies with the local building department or code enforcement agency. These agencies ensure that electrical installations comply with the adopted version of the NEC, which ultimately promotes safety and minimizes the risk of electrical accidents.

106. When using the NEC as a reference, it's important to:
a) Read the entire code book cover to cover
b) Memorize all the articles, sections, and subsections
c) Understand the intent and purpose of the rules
d) Rely on past experience and ignore the actual code language

Answer: c) Understand the intent and purpose of the rules
Explanation: When using the NEC as a reference, it's important to understand the intent and purpose of the rules. This will enable you to apply the code effectively in real-world scenarios and ensure compliance with the code's safety objectives.

107. The NEC uses specific language to indicate the strength of a requirement. Which of the following terms is used to indicate a mandatory requirement?
a) "Shall"
b) "Should"
c) "May"
d) "Can"

Answer: a) "Shall"
Explanation: In the NEC, the term "shall" is used to indicate a mandatory requirement. Understanding the language used in the code is crucial for accurately interpreting its requirements and applying them in real-world scenarios.

108. When you come across an unfamiliar term while navigating the NEC, what should you do?
a) Guess its meaning based on context
b) Consult the definitions in Article 100
c) Assume it has the same meaning as in common usage
d) Ignore it and move on

Answer: b) Consult the definitions in Article 100
Explanation: When encountering an unfamiliar term in the NEC, it's important to consult the definitions in Article 100. This ensures that you are correctly interpreting the term in the context of the code's requirements and objectives.

109. To effectively navigate the NEC and find relevant information, you should:
a) Familiarize yourself with the code's structure and organization
b) Use the index to locate specific requirements
c) Reference the table of contents for a general overview
d) All of the above

Answer: d) All of the above
Explanation: Effectively navigating the NEC requires familiarity with its structure and organization, using the index to locate specific requirements, and referencing the table of contents for a general overview. These strategies will help you quickly and accurately find the information you need in the code.

110. When applying the NEC to a real-world scenario, it is crucial to:
a) Strictly adhere to the letter of the code, regardless of the specific situation
b) Prioritize the spirit and intent of the code, focusing on safety and practicality
c) Only apply the sections that seem most relevant, disregarding the rest
d) Rely on past experience and common sense, rather than the code itself

Answer: b) Prioritize the spirit and intent of the code, focusing on safety and practicality
Explanation: When applying the NEC to real-world scenarios, it is crucial to prioritize the spirit and intent of the code, focusing on safety and practicality. This approach ensures that electrical installations meet the code's objectives and provide a safe and reliable system.

111. One common NEC violation involves the improper installation of ground-fault circuit interrupters (GFCIs). According to the NEC, GFCIs are required for:
a) All residential circuits
b) Only outdoor circuits
c) Specific locations where a potential for ground faults exists
d) Only circuits serving motors and appliances

Answer: c) Specific locations where a potential for ground faults exists
Explanation: The NEC requires GFCIs to be installed in specific locations where there is a potential for ground faults, such as bathrooms, kitchens, garages, and outdoor circuits. Failure to install GFCIs in these locations can result in a code violation and a potentially dangerous situation.

112. A frequent misunderstanding in the NEC involves the proper sizing of electrical conductors. Over-sizing conductors can lead to:
a) Excessive voltage drop
b) Reduced efficiency
c) Increased installation cost without added benefits
d) A decrease in conductor ampacity

Answer: c) Increased installation cost without added benefits
Explanation: Over-sizing conductors can result in increased installation costs without providing any additional benefits. Properly sizing conductors according to the NEC ensures that the electrical system functions safely and efficiently without unnecessary expense.

113. Inadequate working space around electrical equipment is a common NEC violation. According to the NEC, the minimum required working space in front of electrical equipment should be:
a) 24 inches
b) 30 inches
c) 36 inches
d) 48 inches

Answer: c) 36 inches
Explanation: The NEC requires a minimum of 36 inches of working space in front of electrical equipment. This requirement helps ensure that there is enough room for safe access, maintenance, and operation of the equipment.

114. A frequent pitfall encountered when interpreting the NEC is failing to consider:
a) The specific requirements of local jurisdictions
b) The most recent edition of the code
c) The general safety principles outlined in the code
d) The manufacturer's recommendations for equipment installation

Answer: a) The specific requirements of local jurisdictions
Explanation: Failing to consider the specific requirements of local jurisdictions can lead to code violations and non-compliant installations. Local jurisdictions may have additional requirements or modifications to the NEC, and it's important to be familiar with these differences when performing electrical work.

115. One common NEC violation involves the improper use of extension cords. According to the NEC, extension cords should not be used:
a) In lieu of permanent wiring
b) For temporary installations
c) With portable equipment
d) In outdoor locations

Answer: a) In lieu of permanent wiring
Explanation: Using extension cords in place of permanent wiring is a common NEC violation. Extension cords are intended for temporary use and not as a substitute for proper electrical installations. Ensuring that electrical systems are installed according to the NEC requirements helps prevent dangerous situations and code violations.

116. The primary role of an electrical inspector is to:
a) Design electrical systems
b) Install electrical equipment
c) Enforce compliance with the NEC and local regulations
d) Issue permits for electrical work

Answer: c) Enforce compliance with the NEC and local regulations
Explanation: The main responsibility of an electrical inspector is to ensure that electrical work complies with the NEC and local regulations. They review plans, inspect installations, and identify any code violations to help maintain safety and prevent potential hazards.

117. When is it necessary to obtain a permit for electrical work?
a) For any electrical work, regardless of the scope
b) Only for major electrical installations and repairs
c) When the work involves adding or modifying circuits or equipment
d) Only when the work requires a licensed electrician

Answer: c) When the work involves adding or modifying circuits or equipment
Explanation: Permits are generally required for electrical work that involves adding or modifying circuits or equipment. This ensures that the work is reviewed and inspected for compliance with the NEC and local regulations, which helps maintain safety and prevent potential hazards.

118. What is one of the benefits of obtaining a permit for electrical work?
a) It guarantees that the work will be completed on time
b) It ensures that the work is done by a licensed electrician
c) It provides a record of the work for future reference
d) It reduces the cost of the project

Answer: c) It provides a record of the work for future reference
Explanation: Obtaining a permit for electrical work helps create a record of the work that has been done, which can be useful for future reference, such as during property sales or renovations. It also ensures that the work is inspected for compliance with the NEC and local regulations.

119. Electrical inspectors play a crucial role in the permitting process. During the process, they are likely to:
a) Perform the electrical work themselves
b) Review plans and specifications for compliance with the NEC
c) Provide design services for electrical systems
d) Sell electrical equipment and materials

Answer: b) Review plans and specifications for compliance with the NEC
Explanation: Electrical inspectors review plans and specifications during the permitting process to ensure that the proposed work complies with the NEC and local regulations. Their role is to identify any potential issues and ensure that installations are done safely and correctly.

120. Failing to obtain a permit for electrical work can result in:
a) Increased project costs due to fines and penalties
b) A waiver of responsibility for code violations
c) Expedited approval for future projects
d) Automatic approval of the work without inspection

Answer: a) Increased project costs due to fines and penalties
Explanation: Failing to obtain a permit for electrical work can lead to fines and penalties, which can increase the overall cost of the project. Additionally, work completed without a permit may not be inspected for compliance with the NEC and local regulations, potentially resulting in safety hazards and the need for costly corrections.

121. Which of the following study strategies is most effective for mastering the NEC content for the Master Electrician exam?
a) Memorizing the entire NEC word for word
b) Focusing only on the most frequently used articles
c) Developing a deep understanding of the NEC structure and organization
d) Studying only the sections relevant to your specific job

Answer: c) Developing a deep understanding of the NEC structure and organization
Explanation: To effectively prepare for the NEC portion of the Master Electrician exam, it is crucial to develop a deep understanding of the NEC's structure and organization. This will enable you to quickly locate and apply the relevant information during the exam.

122. When preparing for the NEC portion of the Master Electrician exam, it is important to:
a) Rely solely on online forums for information
b) Study only the most recent edition of the NEC
c) Use a variety of resources, including the NEC, practice exams, and study guides
d) Focus on learning the NEC's fine print notes

Answer: c) Use a variety of resources, including the NEC, practice exams, and study guides
Explanation: To effectively prepare for the NEC portion of the Master Electrician exam, using a variety of resources is crucial. Studying the NEC itself, working through practice exams, and utilizing study guides can help ensure a comprehensive understanding of the material.

123. Which of the following is an effective way to familiarize yourself with the NEC language and terminology?
a) Skimming through the entire code quickly
b) Memorizing only the definitions section
c) Reading and working with the code regularly
d) Ignoring the code language and focusing on general concepts

Answer: c) Reading and working with the code regularly
Explanation: Regularly reading and working with the NEC will help you become familiar with its language and terminology. This familiarity will be invaluable during the Master Electrician exam, as it will allow you to understand and apply the code more efficiently.

124. To prepare for the NEC portion of the Master Electrician exam, it is beneficial to:
a) Avoid using practice exams to gauge your progress
b) Focus only on your strengths and ignore your weaknesses
c) Take timed practice exams to simulate the test-taking experience
d) Study the code in isolation without discussing it with others

Answer: c) Take timed practice exams to simulate the test-taking experience
Explanation: Taking timed practice exams can help you simulate the test-taking experience and identify areas where you need to improve. This will also help you become familiar with the types of questions you may encounter on the actual exam.

125. To reinforce your understanding of the NEC, it is helpful to:
a) Study in short, infrequent sessions
b) Avoid using any supplemental resources
c) Discuss code-related topics with peers and colleagues
d) Focus on memorizing rather than understanding the code

Answer: c) Discuss code-related topics with peers and colleagues
Explanation: Discussing code-related topics with peers and colleagues can help reinforce your understanding of the NEC. Sharing knowledge and engaging in conversations about the code can lead to a deeper understanding of the material and help you prepare for the Master Electrician exam.

126. What is the primary danger associated with electrical work that electricians should be most concerned about?
a) Electrical fires
b) Electromagnetic interference
c) Electric shock and electrocution
d) Overheating equipment

Answer: c) Electric shock and electrocution
Explanation: Electric shock and electrocution are the primary dangers associated with electrical work. Electricians should always follow safety guidelines to minimize the risk of injury or death from these hazards.

127. Which of the following is an essential principle of electrical safety?
a) Ignoring grounding and bonding requirements
b) Performing live work whenever possible
c) Using proper personal protective equipment (PPE)
d) Relying on intuition instead of following safety guidelines

Answer: c) Using proper personal protective equipment (PPE)
Explanation: Using proper personal protective equipment (PPE) is an essential principle of electrical safety. PPE helps protect electricians from the risks associated with electrical work, such as electric shock, burns, and arc flash.

128. In the event of an electrical fault, which safety measure helps prevent the flow of electrical current through a person?
a) Short circuit
b) Ground-fault circuit interrupter (GFCI)
c) Proper conductor sizing
d) Overcurrent protection

Answer: b) Ground-fault circuit interrupter (GFCI)
Explanation: A ground-fault circuit interrupter (GFCI) is designed to detect an imbalance in the electrical current flow and quickly shut off the power in the event of a ground fault. This safety measure helps prevent the flow of electrical current through a person, reducing the risk of electric shock and electrocution.

129. Why is it essential to follow safety guidelines when working with electricity?
a) To reduce the risk of accidents and injuries
b) To save time on the job
c) To minimize the use of personal protective equipment
d) To avoid the need for permits and inspections

Answer: a) To reduce the risk of accidents and injuries
Explanation: Following safety guidelines when working with electricity is essential to reduce the risk of accidents and injuries. Adhering to proper safety practices helps protect electricians and others from the hazards associated with electrical work.

130. Which of the following practices is NOT recommended for electrical safety?
a) Performing lockout/tagout procedures when working on electrical equipment
b) Wearing insulated gloves and using insulated tools
c) Assuming that all electrical equipment is energized until proven otherwise
d) Working on live circuits without proper training and authorization

Answer: d) Working on live circuits without proper training and authorization
Explanation: Working on live circuits without proper training and authorization is not recommended for electrical safety. Working on energized equipment increases the risk of electric shock, electrocution, and other hazards. It is crucial to follow established safety procedures and guidelines, such as lockout/tagout, when working with electricity.

131. Which type of personal protective equipment (PPE) is specifically designed to protect electricians from electrical shock when working on energized equipment?
a) Insulated gloves
b) Safety glasses
c) Arc flash suit
d) Steel-toe boots

Answer: a) Insulated gloves
Explanation: Insulated gloves are specifically designed to protect electricians from electrical shock when working on energized equipment. They provide a barrier between the electrician's hands and the live parts, reducing the risk of electric shock and electrocution.

132. What is the primary purpose of wearing safety glasses during electrical work?
a) To protect against ultraviolet radiation
b) To protect the eyes from flying debris and sparks
c) To provide magnification for detailed work
d) To prevent electrical shock

Answer: b) To protect the eyes from flying debris and sparks
Explanation: The primary purpose of wearing safety glasses during electrical work is to protect the eyes from flying debris and sparks. Safety glasses help prevent eye injuries caused by projectiles, such as metal shavings or broken pieces of equipment.

133. Which piece of PPE is essential when working on or near equipment that poses an arc flash hazard?
a) Insulated gloves
b) Safety glasses
c) Arc flash suit
d) Earplugs

Answer: c) Arc flash suit
Explanation: An arc flash suit is essential when working on or near equipment that poses an arc flash hazard. Arc flash suits provide protection from the intense heat, flames, and pressure waves generated during an arc flash incident, reducing the risk of severe burns and other injuries.

134. Why is it crucial to select the correct PPE category when working on electrical equipment?
a) To ensure that the equipment is not damaged
b) To avoid overloading the electrical system
c) To provide the appropriate level of protection for the task
d) To comply with local building codes

Answer: c) To provide the appropriate level of protection for the task
Explanation: Selecting the correct PPE category when working on electrical equipment is crucial to provide the appropriate level of protection for the task. Different PPE categories offer varying degrees of protection, and using the wrong category can result in inadequate protection, increasing the risk of injury.

135. Which of the following is NOT a consideration when selecting PPE for electrical work?
a) The voltage level of the equipment being worked on
b) The color of the PPE
c) The presence of arc flash hazards
d) The type of work being performed

Answer: b) The color of the PPE
Explanation: The color of the PPE is not a consideration when selecting PPE for electrical work. Instead, factors such as the voltage level of the equipment, the presence of arc flash hazards, and the type of work being performed should be considered to ensure the appropriate level of protection.

136. What is the primary purpose of grounding in an electrical system?
a) To minimize voltage drop
b) To provide a path for fault current
c) To reduce electromagnetic interference
d) To increase energy efficiency

Answer: b) To provide a path for fault current
Explanation: The primary purpose of grounding in an electrical system is to provide a path for fault current. This helps ensure that dangerous voltages are safely directed to the ground, reducing the risk of electrical shock and damage to equipment.

137. What is the primary purpose of bonding in an electrical system?
a) To ensure all metallic parts are at the same electrical potential
b) To increase the resistance of the system
c) To provide additional insulation
d) To protect against voltage surges

Answer: a) To ensure all metallic parts are at the same electrical potential
Explanation: The primary purpose of bonding in an electrical system is to ensure all metallic parts are at the same electrical potential. This helps minimize the risk of electrical shock and the potential for arcing between metallic parts, which could result in fires or damage to equipment.

138. According to the NEC, which of the following must be bonded together in an electrical system?
a) All insulated conductors
b) All non-current-carrying metallic parts
c) All current-carrying conductors
d) All underground cables

Answer: b) All non-current-carrying metallic parts
Explanation: According to the NEC, all non-current-carrying metallic parts in an electrical system must be bonded together. This includes enclosures, raceways, and equipment, ensuring that they are all at the same electrical potential, reducing the risk of electrical shock and equipment damage.

139. Which of the following is an example of an effective grounding electrode?
a) Copper water pipe
b) PVC conduit
c) Wooden pole
d) Rubber mat

Answer: a) Copper water pipe
Explanation: A copper water pipe is an example of an effective grounding electrode. Copper water pipes are conductive and have a direct connection to the earth, providing an effective path for fault current to safely dissipate into the ground.

140. In an electrical system, what is the main difference between grounding and bonding?
a) Grounding is for fault protection, while bonding is for overcurrent protection
b) Grounding provides a path for fault current, while bonding ensures all metallic parts are at the same potential
c) Bonding is only required for high-voltage systems, while grounding is required for all systems
d) Grounding is only necessary for equipment, while bonding is required for conductors

Answer: b) Grounding provides a path for fault current, while bonding ensures all metallic parts are at the same potential
Explanation: The main difference between grounding and bonding in an electrical system is their respective purposes. Grounding provides a path for fault current to safely dissipate into the ground, reducing the risk of electrical shock and equipment damage. Bonding ensures all non-current-carrying metallic parts are at the same electrical potential, minimizing the risk of electrical shock and the potential for arcing between metallic parts.

141. Which of the following devices is specifically designed to protect people from electrical shock due to ground faults?
a) Fuse
b) Circuit breaker
c) Ground fault circuit interrupter (GFCI)
d) Surge protector

Answer: c) Ground fault circuit interrupter (GFCI)
Explanation: A ground fault circuit interrupter (GFCI) is specifically designed to protect people from electrical shock due to ground faults. It constantly monitors the current flowing through a circuit and quickly disconnects power if it detects an imbalance, indicating a ground fault.

142. How does a fuse protect an electrical circuit?
a) By reducing voltage drop
b) By disconnecting power when current exceeds a specified limit
c) By providing a path for fault current
d) By eliminating electromagnetic interference

Answer: b) By disconnecting power when current exceeds a specified limit
Explanation: A fuse protects an electrical circuit by disconnecting power when the current exceeds a specified limit. The fuse contains a thin metal wire that melts when the current is too high, breaking the circuit and preventing damage to equipment and reducing the risk of fires.

143. What is the primary function of a circuit breaker in an electrical system?
a) To regulate voltage
b) To protect against power surges
c) To interrupt current flow in case of an overload or short circuit
d) To provide a path for fault current

Answer: c) To interrupt current flow in case of an overload or short circuit
Explanation: The primary function of a circuit breaker in an electrical system is to interrupt current flow in case of an overload or short circuit. When the current exceeds a predetermined level, the circuit breaker opens the circuit, preventing damage to equipment and reducing the risk of fires.

144. In which of the following situations is it most appropriate to use a ground fault circuit interrupter (GFCI)?
a) To protect against voltage surges
b) In outdoor or damp locations where the risk of electrical shock is increased
c) To minimize voltage drop
d) In high-voltage systems to prevent equipment damage

Answer: b) In outdoor or damp locations where the risk of electrical shock is increased
Explanation: A ground fault circuit interrupter (GFCI) is most appropriate for use in outdoor or damp locations where the risk of electrical shock is increased. These devices constantly monitor the current flowing through a circuit and disconnect power if they detect an imbalance, indicating a ground fault.

145. What is the main difference between a fuse and a circuit breaker?
a) Fuses are reusable, while circuit breakers must be replaced after tripping
b) Fuses protect against ground faults, while circuit breakers protect against overcurrent
c) Circuit breakers are reusable, while fuses must be replaced after melting
d) Circuit breakers protect against voltage surges, while fuses protect against short circuits

Answer: c) Circuit breakers are reusable, while fuses must be replaced after melting
Explanation: The main difference between a fuse and a circuit breaker is that circuit breakers are reusable, while fuses must be replaced after melting. Both devices protect electrical circuits by disconnecting power when current exceeds a specified limit, but fuses use a thin metal wire that melts and breaks the circuit, while circuit breakers mechanically open the circuit and can be reset after tripping.

146. Which of the following best describes the purpose of lockout/tagout procedures?
a) To minimize the risk of fire in electrical installations
b) To ensure electrical circuits are properly grounded
c) To prevent unauthorized access to electrical equipment
d) To protect workers from unexpected energization or release of hazardous energy during service and maintenance

Answer: d) To protect workers from unexpected energization or release of hazardous energy during service and maintenance
Explanation: Lockout/tagout procedures are designed to protect workers from unexpected energization or release of hazardous energy during service and maintenance. They involve isolating energy sources, locking or tagging them to prevent accidental re-energizing, and verifying that the equipment is de-energized before work begins.

147. When working with electricity, it is essential to maintain a safe distance from energized conductors. Which of the following factors influences the minimum safe distance?
a) The type of PPE used
b) The voltage of the electrical system
c) The type of tools being used
d) The material of the conductors

Answer: b) The voltage of the electrical system
Explanation: The minimum safe distance when working with electricity depends on the voltage of the electrical system. Higher voltages require greater distances to ensure safety. The safe distance is necessary to prevent electric shock or arc flash incidents.

148. Which of the following tools is specifically designed for safe use in electrical work?
a) Standard pliers
b) Insulated screwdriver
c) Non-sparking hammer
d) Standard wire stripper

Answer: b) Insulated screwdriver
Explanation: Insulated screwdrivers are specifically designed for safe use in electrical work. They have a protective insulation layer on the handle that protects the user from electrical shock when working with energized circuits or components.

149. What is the primary purpose of using non-conductive tools and equipment when working with electricity?
a) To reduce the risk of electrical shock
b) To prevent voltage drops
c) To protect against power surges
d) To minimize electromagnetic interference

Answer: a) To reduce the risk of electrical shock
Explanation: The primary purpose of using non-conductive tools and equipment when working with electricity is to reduce the risk of electrical shock. By using materials that do not conduct electricity, workers can minimize the chances of an electrical current passing through their body if they accidentally contact energized conductors or components.

150. Which of the following is a recommended safety practice for avoiding electrical hazards when working in close proximity to energized circuits?
a) Using only one hand to perform tasks
b) Wearing metal jewelry as a conductor for static electricity
c) Using standard, non-insulated tools
d) Working alone to minimize distractions

Answer: a) Using only one hand to perform tasks
Explanation: Using only one hand to perform tasks when working in close proximity to energized circuits is a recommended safety practice for avoiding electrical hazards. By keeping the other hand away from the work area or in a pocket, workers reduce the risk of creating a path for electrical current to flow through their body, which can cause serious injury or even death.

151. What is the primary cause of arc flash incidents in electrical systems?
a) Insufficient grounding
b) Overloading of circuits
c) Faulty insulation or connections
d) Inadequate ventilation

Answer: c) Faulty insulation or connections. Explanation: Arc flash incidents primarily occur due to faulty insulation or connections in electrical systems. When insulation or connections are damaged or inadequate, it can result in an electrical short circuit, creating a high-energy arc flash that can cause severe burns, hearing damage, or even death.

152. Which of the following best describes the purpose of Ground Fault Circuit Interrupters (GFCIs)?
a) To protect against overcurrent conditions
b) To prevent electrical fires caused by overheating
c) To detect and interrupt ground faults to prevent electric shock
d) To provide backup power during an outage

Answer: c) To detect and interrupt ground faults to prevent electric shock
Explanation: Ground Fault Circuit Interrupters (GFCIs) are designed to detect and interrupt ground faults in order to prevent electric shock. They monitor the flow of current between the hot and neutral conductors and quickly disconnect power if they detect an imbalance, which could indicate a ground fault.

153. In electrical installations, what is the primary purpose of bonding?
a) To create a low-resistance path for fault current to flow
b) To protect circuits from overcurrent
c) To ensure the proper operation of overcurrent protection devices
d) To reduce the risk of electrostatic discharge

Answer: a) To create a low-resistance path for fault current to flow
Explanation: The primary purpose of bonding in electrical installations is to create a low-resistance path for fault current to flow. By providing a path for fault current, bonding ensures that overcurrent protection devices, such as circuit breakers or fuses, operate properly to clear the fault and minimize the risk of electrical hazards like fires or electric shock.

154. What is the most effective way to reduce the risk of electrical fires in residential and commercial buildings?
a) Install surge protectors on all electrical devices
b) Use oversized conductors for all wiring
c) Ensure proper installation and maintenance of electrical systems
d) Install additional grounding rods

Answer: c) Ensure proper installation and maintenance of electrical systems
Explanation: The most effective way to reduce the risk of electrical fires in residential and commercial buildings is to ensure proper installation and maintenance of electrical systems. This includes adhering to NEC requirements, using appropriately sized conductors and overcurrent protection devices, and regularly inspecting and maintaining the electrical system to identify and address potential hazards.

155. Which of the following is a common sign of potential electrical hazards that should be addressed immediately?
a) Circuit breakers that frequently trip
b) The presence of GFCI outlets in kitchens and bathrooms
c) Properly grounded electrical systems
d) The use of insulated tools and equipment

Answer: a) Circuit breakers that frequently trip
Explanation: Circuit breakers that frequently trip are a common sign of potential electrical hazards that should be addressed immediately. Frequent tripping can indicate overloaded circuits, short circuits, or ground faults, which can pose serious risks such as electric shock or fire. Identifying and resolving the underlying cause of the frequent tripping is essential to maintaining a safe electrical system.

156. What is the primary function of an electrical inspector in the permitting process?
a) To design electrical systems
b) To install electrical components
c) To ensure compliance with the NEC and local codes
d) To issue permits for all construction projects

Answer: c) To ensure compliance with the NEC and local codes
Explanation: The primary function of an electrical inspector in the permitting process is to ensure compliance with the National Electrical Code (NEC) and local codes. Electrical inspectors review plans, issue permits, and perform inspections to verify that electrical work meets the necessary safety standards and regulations.

157. Which of the following best describes the importance of obtaining an electrical permit before starting an electrical project?
a) It guarantees that the project will be completed on time
b) It ensures that the work will comply with the NEC and local codes
c) It provides legal protection for the property owner
d) It reduces the cost of the project

Answer: b) It ensures that the work will comply with the NEC and local codes
Explanation: Obtaining an electrical permit before starting an electrical project is important because it ensures that the work will comply with the NEC and local codes. Permits are required for most electrical work to ensure that the installation is safe, meets the necessary regulations, and can be inspected by an electrical inspector.

158. Which of the following actions should be taken if an electrical inspector identifies a code violation during an inspection?
a) Ignore the violation and proceed with the project
b) Request a waiver from the permitting authority
c) Correct the violation and request a re-inspection
d) Report the violation to the local media

Answer: c) Correct the violation and request a re-inspection
Explanation: If an electrical inspector identifies a code violation during an inspection, the appropriate action is to correct the violation and request a re-inspection. Correcting the violation ensures that the electrical work complies with the NEC and local codes, making the installation safe and legal.

159. How does the permitting process help to ensure the safety of electrical installations?
a) It requires all electrical work to be performed by licensed professionals
b) It mandates the use of specific materials and equipment
c) It provides a system of checks and balances, including inspections
d) It forces property owners to take responsibility for electrical safety

Answer: c) It provides a system of checks and balances, including inspections
Explanation: The permitting process helps to ensure the safety of electrical installations by providing a system of checks and balances, including inspections. By requiring permits and inspections for electrical work, the permitting process helps to ensure that installations meet the necessary safety standards and regulations, reducing the risk of hazards such as fires and electric shock.

160. Which of the following is a possible consequence of failing to obtain a permit for electrical work that requires one?
a) Increased energy efficiency
b) Fines, penalties, and potential denial of insurance claims
c) A reduction in property value
d) Approval from the local permitting authority

Answer: b) Fines, penalties, and potential denial of insurance claims
Explanation: Failing to obtain a permit for electrical work that requires one can result in fines, penalties, and potential denial of insurance claims. Performing unpermitted electrical work can also create safety hazards, reduce property value, and lead to legal issues for property owners.

161. Which of the following is a common NEC violation related to the installation of electrical outlets?
a) Installing outlets more than 12 inches above the floor
b) Using outlets with a built-in USB charger
c) Placing outlets too close to water sources
d) Installing outlets in the ceiling

Answer: c) Placing outlets too close to water sources
Explanation: Placing outlets too close to water sources is a common NEC violation. The NEC requires outlets to be installed at a safe distance from water sources, such as sinks, bathtubs, and showers, to minimize the risk of electric shock.

162. What is a frequent mistake made when grounding and bonding electrical systems?
a) Not connecting the grounding electrode conductor to the main bonding jumper
b) Using an oversized grounding electrode conductor
c) Installing too many grounding electrodes
d) Grounding only part of the electrical system

Answer: a) Not connecting the grounding electrode conductor to the main bonding jumper
Explanation: A frequent mistake made when grounding and bonding electrical systems is not connecting the grounding electrode conductor to the main bonding jumper. This connection is essential for providing a low-impedance path to the earth, which helps to prevent electrical hazards, such as shock and fire.

163. Which of the following is a common NEC violation regarding the installation of GFCI protection?
a) Installing GFCI protection on all 15- and 20-ampere, 125-volt receptacles
b) Failing to install GFCI protection in required locations
c) Using GFCI protection in conjunction with AFCI protection
d) Testing GFCI protection devices monthly

Answer: b) Failing to install GFCI protection in required locations
Explanation: Failing to install GFCI protection in required locations is a common NEC violation. The NEC mandates the use of GFCI protection in specific areas, such as kitchens, bathrooms, and outdoors, to minimize the risk of electric shock.

164. What is a common misunderstanding related to the NEC and electrical safety in residential buildings?
a) Believing that the NEC does not apply to residential installations
b) Assuming that all residential electrical work requires a permit
c) Thinking that only licensed electricians can perform residential electrical work
d) Assuming that all electrical systems in residential buildings are grounded

Answer: a) Believing that the NEC does not apply to residential installations
Explanation: A common misunderstanding related to the NEC and electrical safety in residential buildings is believing that the NEC does not apply to residential installations. In fact, the NEC does apply to residential installations and provides safety guidelines and requirements to ensure that electrical systems in homes are safe and reliable.

165. Which of the following is a frequent safety pitfall when working on electrical installations?
a) Failing to de-energize and lockout/tagout the equipment before working on it
b) Using insulated tools and wearing appropriate PPE
c) Working with a qualified partner or observer
d) Regularly inspecting and testing electrical equipment

Answer: a) Failing to de-energize and lockout/tagout the equipment before working on it
Explanation: A frequent safety pitfall when working on electrical installations is failing to de-energize and lockout/tagout the equipment before working on it. This practice is essential for ensuring the safety of workers and minimizing the risk of electrical hazards, such as shock and arc flash.

166. In case of an electrical emergency involving an individual who is in direct contact with a live electrical source, what is the first action you should take?
a) Touch the person to pull them away from the source
b) Use a non-conductive material to separate the person from the source
c) Call an ambulance immediately
d) Attempt to turn off the power source

Answer: d) Attempt to turn off the power source
Explanation: The first action to take in case of an electrical emergency involving a person in direct contact with a live electrical source is to attempt to turn off the power source. This will help to minimize further injury to the person and ensure the safety of others in the area.

167. After encountering someone who has experienced an electrical shock, what is the appropriate first aid measure to take?
a) Apply heat to the affected area
b) Place the person in a sitting position
c) Perform CPR if the person is not breathing
d) Give the person water to drink

Answer: c) Perform CPR if the person is not breathing
Explanation: If a person has experienced an electrical shock and is not breathing, the appropriate first aid measure to take is to perform CPR. CPR can help to maintain oxygen flow to the brain and other vital organs until professional medical help arrives.

168. When should you call 911 in case of an electrical emergency?
a) Only if the person is unconscious
b) If the person has sustained a burn from the electrical contact
c) If the person is experiencing difficulty breathing
d) In any case involving an electrical injury, regardless of severity

Answer: d) In any case involving an electrical injury, regardless of severity
Explanation: In any case involving an electrical injury, regardless of severity, you should call 911. Electrical injuries can have serious internal effects that may not be immediately apparent, and professional medical help should be sought as soon as possible.

169. What type of burn is most commonly associated with electrical injuries?
a) First-degree burn
b) Second-degree burn
c) Third-degree burn
d) Fourth-degree burn

Answer: c) Third-degree burn
Explanation: Third-degree burns are most commonly associated with electrical injuries. These burns are the most severe and involve damage to all layers of the skin, as well as underlying tissues. Immediate medical attention is required for third-degree burns.

170. In the event of an electrical fire, which type of fire extinguisher is most appropriate for use?
a) Class A fire extinguisher
b) Class B fire extinguisher
c) Class C fire extinguisher
d) Class D fire extinguisher

Answer: c) Class C fire extinguisher
Explanation: In the event of an electrical fire, a Class C fire extinguisher is most appropriate for use. Class C fire extinguishers are specifically designed to handle fires involving energized electrical equipment and do not conduct electricity, ensuring the safety of the person using the extinguisher.

171. Why is proper training and qualification essential for electrical workers?
a) To ensure job security and higher pay
b) To create a sense of accomplishment and pride
c) To reduce the risk of accidents and ensure compliance with safety standards
d) To gain recognition from peers and supervisors

Answer: c) To reduce the risk of accidents and ensure compliance with safety standards
Explanation: Proper training and qualification are essential for electrical workers because they help reduce the risk of accidents and ensure compliance with safety standards. Skilled and knowledgeable workers are better equipped to perform their tasks safely and efficiently, thereby protecting themselves and those around them.

172. Which of the following is a significant benefit of obtaining a certification in the electrical field?
a) Increased job opportunities and earning potential
b) Enhanced personal reputation among friends and family
c) Access to exclusive social events and networking opportunities
d) A guaranteed job promotion within the company

Answer: a) Increased job opportunities and earning potential
Explanation: Obtaining a certification in the electrical field can lead to increased job opportunities and earning potential. Certified professionals are often in higher demand and can command better compensation for their expertise and skills.

173. Why is continuing education important for electrical workers?
a) To maintain a competitive edge in the job market
b) To stay informed about the latest industry trends and technological advancements
c) To fulfill a personal desire for lifelong learning
d) To comply with mandatory licensing requirements

Answer: b) To stay informed about the latest industry trends and technological advancements
Explanation: Continuing education is important for electrical workers to stay informed about the latest industry trends and technological advancements. This knowledge helps them stay current with best practices and ensures they can provide the most effective and efficient services to their clients.

174. What role do apprenticeships play in the training and qualification of electrical workers?
a) They provide an opportunity to gain hands-on experience and learn from experienced professionals
b) They serve as a stepping stone to obtaining higher education in the electrical field
c) They are a way for workers to demonstrate their commitment to the industry
d) They offer a chance to network with other professionals in the field

Answer: a) They provide an opportunity to gain hands-on experience and learn from experienced professionals
Explanation: Apprenticeships play a crucial role in the training and qualification of electrical workers by providing an opportunity to gain hands-on experience and learn from experienced professionals. This type of practical training is invaluable in developing the skills necessary to succeed in the electrical industry.

175. How can electrical workers demonstrate their commitment to safety and professionalism?
a) By attending industry conferences and events
b) By obtaining certifications and participating in continuing education programs
c) By displaying their qualifications on their work vehicles and clothing
d) By joining professional organizations and participating in their activities

Answer: b) By obtaining certifications and participating in continuing education programs
Explanation: Electrical workers can demonstrate their commitment to safety and professionalism by obtaining certifications and participating in continuing education programs. These activities show that they are dedicated to staying current with industry best practices and maintaining a high level of competence in their field.

176. What is the primary purpose of using explosion-proof enclosures for electrical equipment in hazardous locations?
a) To prevent ignition of flammable materials by containing sparks and hot surfaces
b) To protect electrical equipment from water and dust ingress
c) To reduce the risk of electrocution for workers in the area
d) To prevent unauthorized access to the electrical equipment

Answer: a) To prevent ignition of flammable materials by containing sparks and hot surfaces
Explanation: Explosion-proof enclosures are designed to prevent ignition of flammable materials by containing sparks and hot surfaces within the enclosure. This ensures that any electrical equipment used in hazardous locations does not inadvertently cause an explosion or fire.

177. Why are isolated power systems used in healthcare facilities?
a) To prevent the spread of infections
b) To minimize the risk of electrical shock to patients and staff
c) To reduce electromagnetic interference with medical equipment
d) To provide backup power in case of a power outage

Answer: b) To minimize the risk of electrical shock to patients and staff
Explanation: Isolated power systems are used in healthcare facilities to minimize the risk of electrical shock to patients and staff. These systems ensure that any fault current is limited and does not pose a significant risk to individuals in the vicinity of the electrical equipment.

178. In outdoor installations, what is an essential consideration when selecting electrical equipment?
a) The distance between the equipment and the nearest building
b) The availability of shade to prevent overheating
c) The equipment's ingress protection (IP) rating
d) The color of the equipment to blend in with the environment

Answer: c) The equipment's ingress protection (IP) rating
Explanation: In outdoor installations, an essential consideration when selecting electrical equipment is its ingress protection (IP) rating. The IP rating indicates the level of protection provided against the ingress of solid objects and liquids, which is crucial for ensuring the equipment's safety and longevity in outdoor environments.

179. What is a critical safety consideration when installing electrical equipment in areas where flammable gases or vapors are present?
a) Ensuring that all equipment is properly grounded and bonded
b) Using equipment specifically designed for use in hazardous locations
c) Installing a ground fault circuit interrupter (GFCI) for each circuit
d) Using non-metallic conduit to avoid sparks caused by friction


Answer: b) Using equipment specifically designed for use in hazardous locations
Explanation: In areas where flammable gases or vapors are present, a critical safety consideration is using equipment specifically designed for use in hazardous locations. This equipment is built to minimize the risk of igniting flammable materials and ensures a safer working environment.

180. What type of lighting is generally preferred for hazardous locations due to its reduced risk of causing ignition?
a) Incandescent lighting
b) Fluorescent lighting
c) High-intensity discharge (HID) lighting
d) Light-emitting diode (LED) lighting


Answer: d) Light-emitting diode (LED) lighting
Explanation: LED lighting is generally preferred for hazardous locations due to its reduced risk of causing ignition. LED lights generate less heat and are less likely to cause sparks compared to other types of lighting, making them a safer choice for environments where flammable materials are present.

181. An electrical contractor is working on a project that involves solving a complex circuit with several resistors connected in parallel and series. What method should they use to find the total resistance of the circuit?
a) Add the values of all resistors in the circuit
b) Multiply the values of all resistors in the circuit
c) Use Ohm's Law to calculate the total resistance
d) Apply a combination of series and parallel resistor formulas

Answer: d) Apply a combination of series and parallel resistor formulas
Explanation: When solving a complex circuit with resistors connected in both parallel and series, the contractor should apply a combination of series and parallel resistor formulas to determine the total resistance.

182. An apprentice electrician is working on a project that requires them to determine the current flowing through a 20-ohm resistor connected to a 120V power source. What is the current in the circuit?
a) 0.6 A
b) 6 A
c) 60 A
d) 2400 A

Answer: b) 6 A
Explanation: To calculate the current in the circuit, the apprentice electrician should use Ohm's Law: I = V / R, where I is the current, V is the voltage, and R is the resistance. In this case, I = 120V / 20 ohms = 6 A.

183. An electrician encounters a circuit with an unknown resistance value. They measure a current of 5 A flowing through the circuit when connected to a 50V power source. What is the resistance of the circuit?
a) 10 ohms
b) 25 ohms
c) 45 ohms
d) 250 ohms

Answer: a) 10 ohms
Explanation: Using Ohm's Law (V = I * R), the electrician can determine the resistance of the circuit by rearranging the formula: R = V / I. In this case, R = 50V / 5A = 10 ohms.

184. A facility manager is concerned about the high power consumption of their lighting system. They measure the voltage across the lights as 240V and the current through the circuit as 20 A. What is the power consumption of the lighting system?
a) 4.8 kW
b) 10 kW
c) 12 kW
d) 4800 W

Answer: a) 4.8 kW
Explanation: The facility manager can use the formula P = V * I to calculate the power consumption of the lighting system. In this case, P = 240V * 20A = 4800 W, which is equal to 4.8 kW.

185. An electrical technician is troubleshooting a motor that is not starting. They measure the voltage across the motor's terminals as 440V and the current as 10 A. If the motor's nameplate indicates a power factor of 0.8, what is the apparent power of the motor?
a) 3.52 kVA
b) 4.4 kVA
c) 5.5 kVA
d) 8.8 kVA


Answer: b) 4.4 kVA
Explanation: To calculate the apparent power (S) of the motor, the technician should use the formula S = V * I. In this case, S = 440V * 10A = 4400 VA, which is equal to 4.4 kVA. Note that the power factor is not needed to calculate the apparent power.

186. An electrical contractor is tasked with installing a new electrical outlet in a residential kitchen. According to the National Electric Code (NEC), what is the minimum required distance between the edge of the kitchen sink and the outlet?
a) 6 inches
b) 12 inches
c) 18 inches
d) 24 inches


Answer: c) 18 inches
Explanation: According to the NEC, a minimum distance of 18 inches is required between the edge of a kitchen sink and an electrical outlet to prevent potential hazards due to water contact.

187. An electrician is working on a commercial building project and needs to install a ground fault circuit interrupter (GFCI) for an outdoor receptacle. What is the minimum height above ground for the outdoor receptacle, as specified by the NEC?
a) 6 inches
b) 12 inches
c) 18 inches
d) 24 inches


Answer: b) 12 inches
Explanation: The NEC requires that outdoor receptacles protected by a GFCI be installed at a minimum height of 12 inches above ground level to prevent potential hazards from moisture or flooding.

188. A contractor is hired to install a new electrical panel in a residential home. According to the NEC, what is the maximum allowable height for the main breaker in the panel?
a) 5 feet
b) 6 feet
c) 6 feet 7 inches
d) 7 feet

Answer: c) 6 feet 7 inches
Explanation: Per the NEC, the maximum height for the main breaker in an electrical panel is 6 feet 7 inches, ensuring that it remains easily accessible for operation.

189. An electrician needs to install a junction box in a commercial building. According to the NEC, what is the minimum required depth for the junction box if it contains 14 AWG conductors?
a) 1.5 inches
b) 2 inches
c) 2.5 inches
d) 3 inches

Answer: a) 1.5 inches
Explanation: The NEC requires a minimum depth of 1.5 inches for junction boxes containing 14 AWG conductors to ensure adequate space for wire connections and prevent potential hazards.

190. A residential homeowner plans to install a swimming pool in their backyard and hires an electrician to ensure compliance with the NEC. What is the minimum required distance between overhead power lines and the swimming pool, as specified by the NEC?
a) 10 feet
b) 15 feet
c) 22.5 feet
d) 25 feet

Answer: c) 22.5 feet
Explanation: According to the NEC, a minimum distance of 22.5 feet is required between overhead power lines and a swimming pool to minimize the risk of electrical hazards.

191. An electrician is working on an energized circuit in a commercial building. According to safety guidelines, which of the following personal protective equipment (PPE) should the electrician wear to prevent electric shock?
a) Leather gloves
b) Insulated gloves
c) Latex gloves
d) Rubber gloves

Answer: b) Insulated gloves
Explanation: Insulated gloves provide the necessary protection against electrical shock when working with energized circuits, as they are specifically designed to protect against electrical hazards.

192. A contractor is installing an electrical panel in a healthcare facility. What type of circuit protection device should be used to minimize the risk of equipment damage and patient harm due to ground faults?
a) Fuse
b) Ground fault circuit interrupter (GFCI)
c) Arc fault circuit interrupter (AFCI)
d) Residual current device (RCD)

Answer: d) Residual current device (RCD)
Explanation: An RCD, also known as a ground fault circuit interrupter (GFCI) in the United States, detects ground faults and disconnects the circuit to prevent equipment damage and protect patients from electrical hazards.

193. An electrician is called to a residential home to troubleshoot a tripping circuit breaker. Upon investigation, they discover a damaged electrical wire. Which of the following is the appropriate course of action to address the damaged wire?
a) Wrap the damaged wire with electrical tape
b) Replace the damaged wire
c) Bypass the damaged wire
d) Install a higher-rated circuit breaker

Answer: b) Replace the damaged wire
Explanation: Replacing the damaged wire is the safest and most appropriate course of action, as it eliminates the potential hazard and ensures the electrical system functions correctly.

194. A team of electricians is working on a large commercial project that requires the de-energization of certain electrical circuits. Which safety procedure should be followed to ensure the protection of the workers and prevent accidental re-energization?
a) Lockout/tagout
b) Personal protective grounding
c) Barricades and signs
d) Safety meetings

Answer: a) Lockout/tagout
Explanation: Lockout/tagout procedures involve securing the circuit disconnecting devices in the off position and applying a tag that warns against re-energization, ensuring the safety of the workers.

195. An electrician is working on a high-voltage installation in an outdoor environment. Which of the following safety measures should be taken to minimize the risk of electrical hazards due to weather conditions?
a) Use non-conductive tools
b) Wear insulated gloves
c) Maintain a safe distance from energized parts
d) Implement a job safety analysis

Answer: a) Use non-conductive tools
Explanation: Using non-conductive tools helps to reduce the risk of electrical hazards when working in outdoor environments where weather conditions, such as rain or humidity, could increase the risk of electric shock.

196. An electrician is working on a project that involves reading blueprints to install electrical systems. What section of the blueprint would provide detailed information on the electrical components and their locations?
a) Architectural drawings
b) Plumbing drawings
c) Electrical drawings
d) Structural drawings

Answer: c) Electrical drawings
Explanation: Electrical drawings provide detailed information on the electrical components and their locations within the building, helping electricians understand the layout and requirements of the electrical system.

197. A project manager is overseeing a large commercial electrical project. Which of the following scheduling methods should they use to ensure the project is completed on time and within budget?
a) Gantt chart
b) Critical path method (CPM)
c) Program evaluation and review technique (PERT)
d) Work breakdown structure (WBS)

Answer: b) Critical path method (CPM)
Explanation: The critical path method (CPM) is a scheduling technique that helps project managers determine the shortest time possible to complete a project by identifying the longest sequence of tasks and their dependencies.

198. An electrician is using a blueprint to determine the location of electrical outlets in a room. What symbol on the blueprint would indicate the location of a standard electrical outlet?
a) A circle with an X in the center
b) A small rectangle with two parallel lines
c) A circle with two parallel lines
d) A small square with an X in the center

Answer: c) A circle with two parallel lines
Explanation: A circle with two parallel lines is the symbol typically used on blueprints to indicate the location of a standard electrical outlet.

199. During a project, the project manager realizes that there is a shortage of qualified electricians, which could delay the project's completion. What should the project manager do to mitigate this risk?
a) Hire additional electricians
b) Request additional funding
c) Reallocate resources from other projects
d) Revise the project schedule

Answer: a) Hire additional electricians
Explanation: Hiring additional electricians will help to address the shortage of qualified workers and prevent delays in the project's completion.

200. A project manager is responsible for overseeing the installation of an electrical system in a new residential development. Which of the following documents should the project manager use to track the project's progress and ensure that it stays on schedule?
a) Scope statement
b) Project budget
c) Project schedule
d) Request for proposal (RFP)

Answer: c) Project schedule. Explanation: A project schedule is a timeline that outlines the start and end dates for each task within the project. This helps the project manager track the project's progress and ensure that it stays on schedule.

201. When working with high voltage electrical systems, what is the minimum safe distance an unqualified person must maintain to avoid electrical hazards?
a) 3 feet
b) 5 feet
c) 10 feet
d) 15 feet

Answer: c) 10 feet
Explanation: An unqualified person should maintain a minimum safe distance of 10 feet from high voltage electrical systems to avoid potential electrical hazards.

202. What type of personal protective equipment (PPE) is crucial for an electrician to wear while working with high voltage systems to prevent arc flash injuries?
a) Safety glasses
b) Insulated gloves
c) Flame-resistant clothing
d) Steel-toed boots

Answer: c) Flame-resistant clothing
Explanation: Flame-resistant clothing is crucial for electricians working with high voltage systems, as it helps protect against the thermal effects of an arc flash, which can cause severe burns.

203. When working on high voltage systems, it is essential to use tools that are specifically designed for this purpose. What characteristic should these tools possess?
a) Insulated handles
b) Non-conductive materials
c) Flame-resistant coating
d) Grounded handles

Answer: b) Non-conductive materials. Explanation: Tools used for working with high voltage systems should be made from non-conductive materials to minimize the risk of electric shock or arc flash.

204. What is the purpose of a safety watch when working on high voltage electrical systems?
a) To monitor the work area for potential hazards
b) To perform the electrical work while the electrician supervises
c) To ensure that safety regulations are being followed
d) To switch off power in case of an emergency

Answer: a) To monitor the work area for potential hazards
Explanation: The safety watch's primary responsibility is to monitor the work area for potential hazards and alert the electrician if any hazards are detected.

205. Before working on a high voltage system, what safety measure should be taken to ensure the system is de-energized and locked out?
a) Use a voltage tester to verify the absence of voltage
b) Turn off the main breaker
c) Disconnect the system from the power source
d) Place a warning sign on the system

Answer: a) Use a voltage tester to verify the absence of voltage
Explanation: Using a voltage tester to verify the absence of voltage ensures that the system is de-energized before work begins. It is essential to follow lockout/tagout procedures to prevent the system from being accidentally re-energized while work is being performed.

206. What is the primary purpose of blueprint reading for electrical professionals?
a) To identify the materials needed for a project
b) To understand the layout and design of electrical systems
c) To determine project deadlines
d) To identify the personnel required for a project

Answer: b) To understand the layout and design of electrical systems
Explanation: The primary purpose of blueprint reading for electrical professionals is to understand the layout and design of electrical systems, which is crucial for proper installation, maintenance, and troubleshooting.

207. Which type of electrical drawing provides a detailed view of a specific electrical component or assembly?
a) Schematic diagram
b) Wiring diagram
c) Single-line diagram
d) Assembly drawing

Answer: d) Assembly drawing
Explanation: Assembly drawings provide a detailed view of a specific electrical component or assembly, including dimensions, materials, and other relevant information.

208. What type of electrical drawing shows the overall layout and arrangement of electrical components within a building or structure?
a) Floor plan
b) Elevation drawing
c) Single-line diagram
d) Wiring diagram

Answer: a) Floor plan
Explanation: Floor plans show the overall layout and arrangement of electrical components within a building or structure, including the location of electrical devices, outlets, and equipment.

209. In blueprint reading, what is the purpose of a legend or symbol key?
a) To indicate the scale of the drawing
b) To provide a list of materials used in the project
c) To explain the meaning of symbols used in the drawing
d) To identify the drawing's author

Answer: c) To explain the meaning of symbols used in the drawing
Explanation: The legend or symbol key in a blueprint explains the meaning of the symbols used in the drawing, allowing the reader to interpret the information accurately.

210. What type of electrical drawing focuses on the connections between components and devices rather than their physical locations?
a) Floor plan
b) Wiring diagram
c) Schematic diagram
d) Elevation drawing

Answer: c) Schematic diagram
Explanation: Schematic diagrams focus on the connections between components and devices, rather than their physical locations, making it easier to understand the overall function and operation of the electrical system.

211. Which symbol is typically used in electrical blueprints to represent a single-pole switch?
a) S
b) SPST
c) SPDT
d) SW

Answer: a) S
Explanation: The symbol "S" is typically used in electrical blueprints to represent a single-pole switch, which is a switch that controls one circuit.

212. In electrical drawings, what does the abbreviation GFCI stand for?
a) Ground Fault Circuit Interrupter
b) Ground Fault Current Interrupter
c) Grounded Feed Circuit Interrupter
d) Ground Faulted Circuit Indicator

Answer: a) Ground Fault Circuit Interrupter
Explanation: The abbreviation GFCI stands for Ground Fault Circuit Interrupter, which is a protective device designed to disconnect a circuit when a potentially hazardous ground fault is detected.

213. What symbol is commonly used to represent a receptacle in electrical blueprints?
a) A circle with an X inside
b) A circle with a T inside
c) A square with a dot inside
d) A square with an R inside

Answer: b) A circle with a T inside
Explanation: A circle with a T inside is commonly used to represent a receptacle in electrical blueprints, indicating where electrical devices can be plugged in.

214. In electrical blueprints, what does the abbreviation EMT represent?
a) Electric Motor Terminal
b) Electrical Metallic Tubing
c) Electrical Management Terminal
d) Electro-Mechanical Transformer

Answer: b) Electrical Metallic Tubing
Explanation: In electrical blueprints, the abbreviation EMT stands for Electrical Metallic Tubing, which is a type of conduit used to protect and route electrical wiring.

215. Which symbol is typically used to represent a junction box in electrical drawings?
a) A solid square
b) A dashed square
c) A solid circle
d) A dashed circle

Answer: c) A solid circle
Explanation: A solid circle is typically used to represent a junction box in electrical drawings, which is an enclosure containing electrical connections and used to protect and organize wiring.

216. What is the primary purpose of a single-line diagram in electrical schematics?
a) To show the physical layout of components
b) To represent the overall electrical system
c) To depict the wiring connections in detail
d) To illustrate the control circuitry of a system

Answer: b) To represent the overall electrical system
Explanation: The primary purpose of a single-line diagram in electrical schematics is to represent the overall electrical system using a simplified, concise format, making it easier to understand the system's layout and organization.

217. What type of information is typically included in multiline diagrams?
a) Only power distribution
b) Only control circuitry
c) Both power distribution and control circuitry
d) Neither power distribution nor control circuitry

Answer: c) Both power distribution and control circuitry
Explanation: Multiline diagrams typically include information about both power distribution and control circuitry, providing a more detailed representation of the electrical system than single-line diagrams.

218. In a single-line diagram, how are electrical components typically represented?
a) Using pictorial symbols
b) Using text labels
c) Using block diagrams
d) Using standardized schematic symbols

Answer: d) Using standardized schematic symbols
Explanation: In a single-line diagram, electrical components are typically represented using standardized schematic symbols, which provide a simplified and easy-to-understand representation of the components and their connections.

219. Which of the following is NOT an advantage of using single-line diagrams?
a) Easier to read and understand
b) Less time-consuming to create
c) Provides detailed wiring information
d) Simplifies complex electrical systems

Answer: c) Provides detailed wiring information
Explanation: Single-line diagrams are advantageous because they are easier to read, understand, and create, as well as simplifying complex electrical systems. However, they do not provide detailed wiring information, which is typically found in multiline diagrams.

220. In multiline diagrams, what do solid lines usually represent?
a) Physical connections between components
b) Electrical connections between components
c) Control signals between components
d) Mechanical linkages between components

Answer: b) Electrical connections between components
Explanation: In multiline diagrams, solid lines usually represent electrical connections between components, showing the path of current flow and wiring connections within the system.

221. What is the primary purpose of a wiring diagram in electrical projects?
a) To outline the physical layout of an electrical system
b) To indicate the types and ratings of electrical components
c) To provide detailed information about electrical connections
d) To schedule the installation of electrical components

Answer: c) To provide detailed information about electrical connections
Explanation: The primary purpose of a wiring diagram is to provide detailed information about the electrical connections within a system, including how components are interconnected and the routing of wires.

222. What is the main objective of a panel schedule in an electrical project?
a) To track the progress of the project
b) To organize and document circuit information
c) To schedule maintenance tasks for electrical panels
d) To create a budget for electrical components

Answer: b) To organize and document circuit information
Explanation: The main objective of a panel schedule is to organize and document circuit information related to an electrical panel, such as breaker sizes, load calculations, and the purpose of each circuit.

223. What information is typically included in a panel schedule?
a) Cable types and sizes
b) Circuit breaker ratings and circuit descriptions
c) Detailed wiring diagrams for each circuit
d) Installation instructions for electrical devices

Answer: b) Circuit breaker ratings and circuit descriptions
Explanation: A panel schedule typically includes information about circuit breaker ratings and circuit descriptions, providing an organized and easily accessible record of the electrical panel's components and their functions.

224. Which of the following is NOT a common element found in wiring diagrams?
a) Standardized schematic symbols
b) Wire color codes
c) Load calculations for each circuit
d) Component identification labels

Answer: c) Load calculations for each circuit
Explanation: Load calculations for each circuit are typically not included in wiring diagrams. Wiring diagrams focus on providing detailed information about electrical connections, using standardized schematic symbols, wire color codes, and component identification labels.

225. In a wiring diagram, what do dashed lines typically represent?
a) The physical layout of components
b) Electrical connections between components
c) Control or communication connections between components
d) Mechanical linkages between components

Answer: c) Control or communication connections between components
Explanation: In a wiring diagram, dashed lines typically represent control or communication connections between components, distinguishing them from solid lines, which represent electrical connections.

226. Why is coordination with other trades essential when planning electrical installations?
a) To ensure all tradespeople follow the same schedule
b) To minimize costs and reduce project delays
c) To ensure proper installation of electrical equipment
d) To prevent conflicts and interferences between systems

Answer: d) To prevent conflicts and interferences between systems
Explanation: Coordination with other trades, such as HVAC or plumbing, is essential to prevent conflicts and interferences between systems, ensuring that each system is installed correctly and functions as intended.

227. What is a common issue that may arise if coordination with other trades is not properly conducted during electrical installations?
a) Increased project costs due to rework and modifications
b) Miscommunication between project managers
c) Difficulty obtaining permits for the project
d) Unnecessary use of electrical components

Answer: a) Increased project costs due to rework and modifications
Explanation: A lack of coordination with other trades can lead to conflicts and interferences between systems, often resulting in increased project costs due to rework and modifications required to correct the issues.

228. When reviewing blueprints, which of the following actions should an electrician take to ensure proper coordination with other trades?
a) Consult the project manager for guidance
b) Review other trade drawings and discuss potential conflicts
c) Focus only on the electrical drawings and trust other trades to coordinate
d) Communicate only with the electrical inspector

Answer: b) Review other trade drawings and discuss potential conflicts
Explanation: To ensure proper coordination with other trades, electricians should review other trade drawings and discuss potential conflicts with relevant professionals, working together to find solutions and prevent issues during installation.

229. How can proper coordination with other trades contribute to overall project success?
a) By reducing the need for electrical inspections
b) By ensuring a faster installation process
c) By enhancing the safety and functionality of the completed project
d) By eliminating the need for permits

Answer: c) By enhancing the safety and functionality of the completed project
Explanation: Proper coordination with other trades can enhance the safety and functionality of the completed project by preventing conflicts and interferences between systems, ensuring that each system is installed correctly and operates as intended.

230. Which of the following is NOT a benefit of coordinating with other trades during an electrical installation project?
a) Minimized risk of damage to other systems
b) Improved communication among project team members
c) Faster permitting process
d) Reduced likelihood of project delays

Answer: c) Faster permitting process
Explanation: While coordinating with other trades can provide many benefits, it does not directly impact the permitting process. The primary benefits of coordination include minimizing the risk of damage to other systems, improving communication among project team members, and reducing the likelihood of project delays.

231. Why is project management important in the electrical field?
a) To ensure efficient use of materials and resources
b) To effectively coordinate with other trades
c) To meet project deadlines and stay within budget
d) All of the above

Answer: d) All of the above
Explanation: Project management is important in the electrical field because it ensures efficient use of materials and resources, effective coordination with other trades, and helps to meet project deadlines and stay within budget.

232. Which of the following is NOT a key concept in project management for the electrical field?
a) Scope management
b) Time management
c) Cost management
d) Product marketing

Answer: d) Product marketing
Explanation: Product marketing is not a key concept in project management for the electrical field. Key concepts include scope management, time management, and cost management.

233. What is the primary role of an electrical project manager?
a) To supervise and direct the work of electricians on a project
b) To procure materials and negotiate contracts
c) To design electrical systems and create blueprints
d) To ensure compliance with safety regulations

Answer: a) To supervise and direct the work of electricians on a project
Explanation: The primary role of an electrical project manager is to supervise and direct the work of electricians on a project, ensuring that work is completed on time, within budget, and in accordance with safety regulations and industry standards.

234. In the context of electrical project management, what does "scope management" entail?
a) Allocating resources and managing costs
b) Defining and controlling the work to be performed on a project
c) Ensuring that deadlines are met and schedules are maintained
d) Supervising the installation and maintenance of electrical equipment

Answer: b) Defining and controlling the work to be performed on a project
Explanation: Scope management in electrical project management entails defining and controlling the work to be performed on a project, ensuring that the project's goals and objectives are clearly understood and that any changes are managed appropriately.

235. What is the main purpose of cost management in electrical project management?
a) To determine the profitability of a project
b) To estimate and control project expenses
c) To ensure that materials and equipment are purchased at the lowest price
d) To manage the salaries and benefits of project team members

Answer: b) To estimate and control project expenses
Explanation: The main purpose of cost management in electrical project management is to estimate and control project expenses, ensuring that the project stays within budget and that resources are used efficiently.

236. During which stage of project management is the feasibility and value of the project determined?
a) Initiation
b) Planning
c) Execution
d) Closure

Answer: a) Initiation
Explanation: The initiation stage is when the feasibility and value of the project are determined. This stage involves identifying the project's goals and objectives, assessing its viability, and obtaining approval to proceed.

237. What is the primary focus of the planning stage in project management?
a) Defining the project's scope and objectives
b) Allocating resources and developing a detailed schedule
c) Managing and controlling project activities
d) Reviewing and evaluating the project's performance

Answer: b) Allocating resources and developing a detailed schedule
Explanation: The planning stage of project management primarily focuses on allocating resources, developing a detailed schedule, and creating a project management plan that guides the project through its execution.

238. Which stage of project management involves the actual implementation of the project plan?
a) Initiation
b) Planning
c) Execution
d) Monitoring and control

Answer: c) Execution
Explanation: The execution stage of project management involves the actual implementation of the project plan. This stage includes carrying out the tasks, managing resources, and ensuring that the project progresses according to the plan.

239. What is the primary purpose of the monitoring and control stage of project management?
a) To review and approve the project plan
b) To ensure that project activities align with the plan and to make adjustments as needed
c) To assess project performance and identify areas for improvement
d) To finalize the project and hand it over to the client

Answer: b) To ensure that project activities align with the plan and to make adjustments as needed
Explanation: The primary purpose of the monitoring and control stage of project management is to ensure that project activities align with the plan and to make adjustments as needed. This stage involves tracking progress, evaluating performance, and implementing corrective actions to keep the project on track.

240. What is the main objective of the closure stage in project management?
a) To finalize and document all project activities
b) To assess the project's success and identify lessons learned
c) To hand over the completed project to the client
d) All of the above

Answer: d) All of the above
Explanation: The main objective of the closure stage in project management is to finalize and document all project activities, assess the project's success and identify lessons learned, and hand over the completed project to the client. This stage ensures that all aspects of the project are properly closed and that the project's objectives have been met.

241. What is a key element of a well-defined project scope in an electrical project?
a) A comprehensive list of project objectives
b) A detailed description of the project's deliverables and boundaries
c) A plan for resource allocation
d) A timeline for project completion

Answer: b) A detailed description of the project's deliverables and boundaries
Explanation: A key element of a well-defined project scope is a detailed description of the project's deliverables and boundaries. This description helps to clarify what is included and excluded from the project, ensuring all stakeholders have a clear understanding of the work to be completed.

242. Which of the following is crucial for setting clear objectives in an electrical project?
a) Ensuring objectives are achievable and measurable
b) Making objectives as broad as possible
c) Setting objectives without considering available resources
d) Defining objectives without input from stakeholders

Answer: a) Ensuring objectives are achievable and measurable
Explanation: Ensuring objectives are achievable and measurable is crucial for setting clear objectives in an electrical project. This allows project managers to effectively track progress and make adjustments as needed to ensure project success.

243. What is the primary purpose of defining a project's scope and objectives?
a) To develop a detailed project schedule
b) To provide a framework for decision-making and resource allocation
c) To identify potential risks and challenges
d) To create a communication plan

Answer: b) To provide a framework for decision-making and resource allocation
Explanation: The primary purpose of defining a project's scope and objectives is to provide a framework for decision-making and resource allocation. This ensures that the project stays on track and that resources are used efficiently to achieve the project's goals.

244. Which of the following is a common mistake when defining project scope in an electrical project?
a) Including too much detail
b) Failing to consider stakeholder input
c) Overestimating available resources
d) Setting unrealistic objectives

Answer: b) Failing to consider stakeholder input
Explanation: Failing to consider stakeholder input is a common mistake when defining project scope in an electrical project. Stakeholder input is essential for ensuring that the project's scope accurately reflects the needs and expectations of all parties involved, which ultimately contributes to project success.

245. Which of the following is an example of a well-defined objective for an electrical project?
a) Complete the project within budget
b) Install 50 energy-efficient lighting fixtures by the end of the month
c) Improve the electrical system's efficiency
d) Meet all deadlines

Answer: b) Install 50 energy-efficient lighting fixtures by the end of the month
Explanation: A well-defined objective for an electrical project should be specific, measurable, achievable, relevant, and time-bound (SMART). Option b) is an example of a well-defined objective, as it specifies a clear and measurable goal with a set deadline.

246. What is the primary purpose of budgeting in an electrical project?
a) To ensure that the project is completed on time
b) To track progress and make adjustments as needed
c) To allocate resources efficiently and control costs
d) To identify potential risks and challenges

Answer: c) To allocate resources efficiently and control costs
Explanation: The primary purpose of budgeting in an electrical project is to allocate resources efficiently and control costs. This ensures that the project stays within the financial constraints while maximizing the value delivered.

247. Which of the following factors should be considered when allocating resources for an electrical project?
a) The cost of labor, materials, and equipment
b) The preferences of stakeholders
c) The project's scope and objectives
d) All of the above

Answer: d) All of the above
Explanation: When allocating resources for an electrical project, it is essential to consider the cost of labor, materials, and equipment, the preferences of stakeholders, and the project's scope and objectives. This ensures that resources are allocated effectively and contribute to the overall success of the project.

248. What is a potential consequence of inadequate budgeting in an electrical project?
a) Increased efficiency
b) Cost overruns
c) Reduced communication among team members
d) Faster project completion

Answer: b) Cost overruns
Explanation: Inadequate budgeting in an electrical project can lead to cost overruns, as resources may not be allocated efficiently, and unexpected expenses may arise. This can have a negative impact on the project's overall success and stakeholder satisfaction.

249. Which of the following resource allocation strategies can help reduce labor costs in an electrical project?
a) Hiring more experienced workers
b) Cross-training team members to perform multiple tasks
c) Investing in high-quality equipment
d) Reducing the project's scope

Answer: b) Cross-training team members to perform multiple tasks
Explanation: Cross-training team members to perform multiple tasks can help reduce labor costs in an electrical project. This approach enables workers to fill in for one another as needed, leading to increased efficiency and flexibility in managing the workforce.

250. In an electrical project, how can a project manager ensure that the budget for materials is used effectively?
a) By purchasing the cheapest materials available
b) By selecting materials based on stakeholder preferences
c) By accurately estimating material requirements and monitoring usage
d) By focusing only on the materials needed for the initial phase of the project

Answer: c) By accurately estimating material requirements and monitoring usage
Explanation: To ensure that the budget for materials is used effectively in an electrical project, a project manager should accurately estimate material requirements and monitor usage throughout the project. This approach helps to prevent waste and ensure that materials are used efficiently to achieve the project's objectives.

251. What is the primary purpose of a Gantt chart in project management?
a) To identify risks and challenges
b) To track budget and expenses
c) To visualize the project schedule and progress
d) To allocate resources efficiently

Answer: c) To visualize the project schedule and progress
Explanation: A Gantt chart is a visual representation of the project schedule, showing tasks, durations, dependencies, and progress. It helps project managers track the project's progress and make adjustments as needed.

252. What is the critical path in a project schedule?
a) The sequence of tasks with the least amount of flexibility
b) The shortest path through the project
c) The sequence of tasks that must be completed on time to avoid delaying the project
d) The path with the most significant risks and challenges

Answer: c) The sequence of tasks that must be completed on time to avoid delaying the project
Explanation: The critical path is the sequence of tasks that must be completed on time to avoid delaying the project. It represents the longest duration path through the project, taking into account task dependencies and durations.

253. Which scheduling technique helps identify the early start and early finish times for project tasks?
a) Gantt chart
b) Resource leveling
c) Critical chain method
d) Forward pass

Answer: d) Forward pass
Explanation: The forward pass is a scheduling technique used to calculate the early start and early finish times for each task in the project schedule. It helps identify critical tasks and slack time, enabling better timeline management.

254. What is the primary benefit of using the critical path method (CPM) in project scheduling?
a) It simplifies the project schedule
b) It reduces the need for resource allocation
c) It identifies the tasks that directly impact the project completion date
d) It eliminates the need for monitoring and controlling the project

Answer: c) It identifies the tasks that directly impact the project completion date
Explanation: The critical path method (CPM) is a project scheduling technique that identifies the tasks that directly impact the project completion date. By focusing on these tasks, project managers can better manage the project timeline and mitigate potential delays.

255. How can a project manager use float or slack time in project scheduling?
a) To allocate additional resources to tasks on the critical path
b) To identify tasks that can be delayed without affecting the project completion date
c) To reduce the overall duration of the project
d) To establish a more aggressive timeline

Answer: b) To identify tasks that can be delayed without affecting the project completion date
Explanation: Float or slack time refers to the amount of time a task can be delayed without affecting the project completion date. Project managers can use this information to prioritize tasks and manage resources more effectively, focusing on tasks with little or no slack time.

256. What is the primary purpose of risk management in electrical projects?
a) To eliminate all risks associated with the project
b) To identify and mitigate risks that could negatively impact project success
c) To transfer all risks to other stakeholders
d) To reduce project costs

Answer: b) To identify and mitigate risks that could negatively impact project success
Explanation: Risk management in electrical projects aims to identify and mitigate risks that could negatively impact project success. By proactively addressing risks, project managers can increase the likelihood of a successful project outcome.

257. Which of the following is NOT a step in the risk management process?
a) Risk identification
b) Risk assessment
c) Risk elimination
d) Risk response planning

Answer: c) Risk elimination
Explanation: Risk elimination is not a step in the risk management process. While it's ideal to eliminate risks entirely, this is often unrealistic. The risk management process involves risk identification, risk assessment, risk response planning, and risk monitoring and control.

258. Which risk assessment technique is best suited for quantifying the potential impact of risks on a project's objectives?
a) Qualitative risk analysis
b) Quantitative risk analysis
c) Risk avoidance
d) Risk transfer

Answer: b) Quantitative risk analysis
Explanation: Quantitative risk analysis is a risk assessment technique that uses numerical data to quantify the potential impact of risks on a project's objectives. This helps project managers prioritize risks and allocate resources accordingly.

259. Which risk response strategy involves reallocating project resources to reduce the impact of a specific risk?
a) Risk avoidance
b) Risk mitigation
c) Risk transfer
d) Risk acceptance

Answer: b) Risk mitigation
Explanation: Risk mitigation involves taking action to reduce the probability or impact of a risk. This can include reallocating project resources, implementing additional safety measures, or modifying project plans to address the risk.

260. In the context of risk management, what is the purpose of a contingency plan?
a) To identify potential risks
b) To assess the impact of risks on project objectives
c) To provide a predefined course of action in case a risk event occurs
d) To transfer risks to other stakeholders

Answer: c) To provide a predefined course of action in case a risk event occurs
Explanation: A contingency plan is a predefined course of action designed to address a specific risk event if it occurs. By having contingency plans in place, project managers can respond quickly and effectively to minimize the impact of the risk event on the project.

261. What is the primary goal of effective communication in managing electrical projects?
a) To impress stakeholders with technical jargon
b) To ensure all team members understand their roles and responsibilities
c) To minimize the need for face-to-face meetings
d) To avoid discussing potential problems

Answer: b) To ensure all team members understand their roles and responsibilities
Explanation: The primary goal of effective communication in managing electrical projects is to ensure that all team members understand their roles and responsibilities. Clear communication helps to avoid misunderstandings and ensure that tasks are completed on time and as expected.

262. Which of the following is NOT a characteristic of an effective leader in electrical project management?
a) The ability to motivate and inspire team members
b) A clear understanding of project goals and objectives
c) A tendency to avoid conflict at all costs
d) Good problem-solving and decision-making skills

Answer: c) A tendency to avoid conflict at all costs
Explanation: An effective leader in electrical project management should not avoid conflict at all costs. Instead, they should address conflicts and challenges proactively, working to resolve them in a constructive manner that benefits the project and the team.

263. What is the primary purpose of active listening in the context of project management?
a) To show dominance and authority in a conversation
b) To gather information and understand others' perspectives
c) To multitask and plan the next response while others are speaking
d) To minimize the need for follow-up conversations

Answer: b) To gather information and understand others' perspectives
Explanation: Active listening in the context of project management is crucial for gathering information and understanding others' perspectives. This skill helps project managers make informed decisions, address concerns, and promote a collaborative work environment.

264. Which communication style is most effective for fostering collaboration and teamwork in electrical projects?
a) Aggressive communication
b) Passive communication
c) Assertive communication
d) Passive-aggressive communication

Answer: c) Assertive communication
Explanation: Assertive communication is the most effective style for fostering collaboration and teamwork in electrical projects. This approach involves expressing thoughts and feelings openly, honestly, and respectfully, ensuring that all parties feel heard and valued.

265. How can a project manager demonstrate effective leadership when delegating tasks in an electrical project?
a) By providing clear instructions and expectations
b) By micromanaging every aspect of the task
c) By assigning tasks without any explanation
d) By delegating only menial tasks to team members

Answer: a) By providing clear instructions and expectations
Explanation: Demonstrating effective leadership when delegating tasks involves providing clear instructions and expectations. This ensures that team members understand their responsibilities and have the necessary information to complete their tasks successfully.

266. What is the primary purpose of implementing quality control measures in electrical projects?
a) To reduce costs by cutting corners
b) To prevent potential rework and warranty claims
c) To create unnecessary documentation
d) To shift the blame to other team members

Answer: b) To prevent potential rework and warranty claims
Explanation: The primary purpose of implementing quality control measures in electrical projects is to prevent potential rework and warranty claims. Ensuring high-quality work helps to maintain project timelines, reduce costs, and improve overall customer satisfaction.

267. Which of the following is NOT a key component of an effective quality assurance plan for electrical projects?
a) Regular inspections and testing of electrical installations
b) Comprehensive documentation of project processes and results
c) Cutting corners to save time and resources
d) Continual improvement and learning from past experiences

Answer: c) Cutting corners to save time and resources
Explanation: Cutting corners to save time and resources is not a key component of an effective quality assurance plan. Instead, quality assurance plans should focus on maintaining high-quality work, even if it requires additional time and resources, to avoid potential rework and warranty claims.

268. What is the main difference between quality control and quality assurance in electrical projects?
a) Quality control focuses on preventing defects, while quality assurance focuses on detecting defects
b) Quality control focuses on detecting defects, while quality assurance focuses on preventing defects
c) Quality control is only applicable during project execution, while quality assurance is only applicable during project planning
d) Quality control and quality assurance are interchangeable terms

Answer: b) Quality control focuses on detecting defects, while quality assurance focuses on preventing defects

Explanation: Quality control focuses on detecting defects through inspections and testing, while quality assurance focuses on preventing defects by establishing and implementing processes and procedures that ensure high-quality work.

269. In the context of electrical projects, what is the primary benefit of using a quality management system (QMS)?
a) A QMS eliminates the need for inspections and testing
b) A QMS guarantees that all projects will be completed ahead of schedule
c) A QMS provides a structured approach to managing and improving quality
d) A QMS ensures that no mistakes will ever be made on a project

Answer: c) A QMS provides a structured approach to managing and improving quality
Explanation: A quality management system (QMS) provides a structured approach to managing and improving quality in electrical projects. By implementing a QMS, project managers can establish processes and procedures that help to ensure high-quality work and continuous improvement.

270. What is the primary role of a quality control inspector in an electrical project?
a) To determine project scope and objectives
b) To supervise and manage project team members
c) To perform inspections and tests to verify compliance with codes and specifications
d) To approve change orders and manage project budgets

Answer: c) To perform inspections and tests to verify compliance with codes and specifications
Explanation: The primary role of a quality control inspector in an electrical project is to perform inspections and tests to verify compliance with codes and specifications. This helps to ensure that the project meets the required quality standards and reduces the risk of rework or warranty claims.

271. What is the primary objective of the project closure phase in an electrical project?
a) To start planning for the next project
b) To finalize all project-related tasks and documentation
c) To assign blame for any issues that arose during the project
d) To request additional funding for future projects

Answer: b) To finalize all project-related tasks and documentation
Explanation: The primary objective of the project closure phase in an electrical project is to finalize all project-related tasks and documentation. This includes ensuring that all work has been completed, obtaining final approvals and inspections, and archiving project documentation for future reference.

272. What is the primary reason for conducting a lessons learned session at the end of an electrical project?
a) To celebrate the project's successes
b) To identify areas for improvement and apply them to future projects
c) To provide a forum for team members to complain about the project
d) To decide which team members will be promoted

Answer: b) To identify areas for improvement and apply them to future projects
Explanation: The primary reason for conducting a lessons learned session at the end of an electrical project is to identify areas for improvement and apply them to future projects. This process helps the project team learn from their experiences and continuously improve their project management skills.

273. Which of the following is NOT a critical component of project closure documentation in electrical projects?
a) Final inspection reports and certificates of occupancy
b) A detailed list of every team member's personal strengths and weaknesses
c) As-built drawings and updated schematics
d) Warranty and maintenance information for installed equipment

Answer: b) A detailed list of every team member's personal strengths and weaknesses
Explanation: While it is important to evaluate team performance, a detailed list of every team member's personal strengths and weaknesses is not a critical component of project closure documentation. Instead, focus on the documentation that pertains directly to the project's completion, such as inspection reports, as-built drawings, and warranty information.

274. What is the primary purpose of obtaining a certificate of occupancy at the end of an electrical project?
a) To indicate that the project has been completed according to code requirements and is safe for occupancy
b) To provide proof of insurance coverage for the building
c) To serve as a marketing tool for attracting tenants
d) To verify that all project payments have been made

Answer: a) To indicate that the project has been completed according to code requirements and is safe for occupancy
Explanation: The primary purpose of obtaining a certificate of occupancy at the end of an electrical project is to indicate that the project has been completed according to code requirements and is safe for occupancy. This document is issued by the local building department and is required before the building can be legally occupied.

275. Why is it important to archive all project documentation after the completion of an electrical project?
a) To comply with legal and regulatory requirements
b) To provide a historical record for reference in future projects
c) To ensure that all team members receive credit for their work
d) Both a) and b)

Answer: d) Both a) and b)
Explanation: Archiving all project documentation after the completion of an electrical project is important for two main reasons: to comply with legal and regulatory requirements, and to provide a historical record for reference in future projects. This documentation can be invaluable for troubleshooting, maintenance, and future project planning.

276. An electrical contractor is working on a commercial building project. They need to calculate the size of the feeder conductor for a 3-phase, 120/208-volt panelboard. The panelboard's total load is 96 kVA. Given the following conductor sizes and their respective ampacity, which conductor size should the contractor choose?
a) 1/0 AWG (230 A)
b) 2/0 AWG (255 A)
c) 3/0 AWG (285 A)
d) 4/0 AWG (310 A)

Answer: b) 2/0 AWG (255 A)
Explanation: First, calculate the total current: I = (96,000 VA) / (208 V x √3) ≈ 267 A. The contractor should choose the smallest conductor size with an ampacity greater than the calculated current, which is 2/0 AWG with an ampacity of 255 A.

277. An electrician needs to calculate the voltage drop for a 240-volt, single-phase circuit with a current of 50 A and a conductor length of 150 feet. The circuit uses copper conductors with a resistance of 0.2 ohms per 1000 feet. What is the voltage drop?
a) 3 V
b) 6 V
c) 9 V
d) 12 V

Answer: c) 9 V
Explanation: Voltage drop = 2 x conductor length x current x conductor resistance / 1000. Using the given values, voltage drop = 2 x 150 x 50 x 0.2 / 1000 = 9 V.

278. An electrical installation requires a 480-volt, 3-phase motor with a full-load current of 45 A. If the short-circuit current at the motor terminals is 6,000 A, what is the minimum required interrupting rating for the motor's circuit breaker?
a) 6,000 A
b) 12,000 A
c) 18,000 A
d) 24,000 A

Answer: a) 6,000 A
Explanation: The minimum required interrupting rating for the motor's circuit breaker should be equal to or greater than the short-circuit current at the motor terminals, which is 6,000 A.

279. An electrician needs to calculate the maximum number of 12 AWG THHN conductors that can be installed in a 3/4-inch EMT conduit, given a maximum conduit fill of 40%. What is the maximum number of conductors allowed?
a) 9
b) 12
c) 16
d) 20

Answer: c) 16
Explanation: First, determine the area of a 3/4-inch EMT conduit (0.508 in²). Multiply this area by the maximum conduit fill (0.508 in² x 0.40 = 0.2032 in²). Then, divide the result by the approximate area of a single 12 AWG THHN conductor (0.0133 in²): 0.2032 in² / 0.0133 in² ≈ 15.27. The maximum number of conductors allowed is 16.

280. An electrician is installing a 240-volt, single-phase, 5-horsepower motor with a power factor of 0.85. What is the minimum required ampacity for the motor's branch-circuit conductors?
a) 15 A
b) 20 A
c) 25 A
d) 30 A

Answer: b) 20 A

281. Electrical calculations are essential for professionals in the electrical field because they help to:
a) Create aesthetic designs for electrical systems
b) Ensure the safety and efficiency of electrical systems
c) Communicate with clients about their electrical preferences
d) Determine the colors of wires to be used in installations

Answer: b) Ensure the safety and efficiency of electrical systems
Explanation: Electrical calculations are crucial for designing safe and efficient electrical systems, which helps prevent electrical hazards, such as fires, and ensures optimal system performance.

282. One of the key concepts covered in the electrical calculations chapter is the calculation of:
a) Square footage of a building
b) Voltage drop in a conductor
c) Number of outlets needed in a room
d) Length of conduit required for a project

Answer: b) Voltage drop in a conductor
Explanation: Voltage drop calculations are essential in the electrical field to ensure that electrical systems operate within acceptable voltage limits and to prevent potential issues such as overheating or equipment malfunction.

283. In electrical calculations, Ohm's Law is used to determine the relationship between:
a) Voltage, current, and resistance
b) Current, resistance, and power
c) Voltage, resistance, and power
d) Voltage, current, and power

Answer: a) Voltage, current, and resistance
Explanation: Ohm's Law is a fundamental principle in electrical calculations, describing the relationship between voltage (V), current (I), and resistance (R) in an electrical circuit. The formula is $V = IR$.

284. To calculate the current in a circuit with a known voltage and resistance, an electrician would use the formula:
a) $I = V / R$
b) $V = IR$
c) $P = IV$
d) $R = V / I$

Answer: a) $I = V / R$. Explanation: The formula $I = V / R$ is derived from Ohm's Law ($V = IR$) and is used to calculate the current in a circuit when the voltage and resistance are known.

285. When calculating the power consumed by an electrical device, an electrician uses the formula:
a) $P = I^2R$
b) $P = V / R$
c) $P = IV$
d) $P = V^2 / R$

Answer: c) P = IV
Explanation: The formula P = IV is used to calculate the power (P) consumed by an electrical device, where I represents the current and V represents the voltage.

286. Ohm's Law can be applied to calculate the current in a circuit when the voltage is 120 V and the resistance is 40 ohms. What is the current in the circuit?
a) 1 A
b) 3 A
c) 4 A
d) 8 A

Answer: b) 3 A
Explanation: Using Ohm's Law (I = V / R), we can calculate the current (I) by dividing the voltage (V) by the resistance (R). In this case, I = 120 V / 40 ohms = 3 A.

287. An electrical device consumes 240 watts of power and operates at a voltage of 120 V. What is the current flowing through the device?
a) 0.5 A
b) 1 A
c) 2 A
d) 3 A

Answer: c) 2 A
Explanation: Using the power formula P = IV, we can solve for the current (I) by dividing the power (P) by the voltage (V). In this case, I = 240 W / 120 V = 2 A.

288. If a 10-ohm resistor has a current of 2 A flowing through it, what is the voltage across the resistor?
a) 5 V
b) 10 V
c) 15 V
d) 20 V

Answer: d) 20 V
Explanation: Using Ohm's Law (V = IR), we can calculate the voltage (V) by multiplying the current (I) by the resistance (R). In this case, V = 2 A × 10 ohms = 20 V.

289. In an electrical circuit, the current is 4 A, and the resistance is 25 ohms. What is the power dissipated by the circuit?
a) 100 W
b) 200 W
c) 400 W
d) 800 W

Answer: c) 400 W
Explanation: First, we need to find the voltage using Ohm's Law (V = IR). In this case, V = 4 A × 25 ohms = 100 V. Then, we can use the power formula P = IV to calculate the power. P = 4 A × 100 V = 400 W.

290. An electrical motor has a power rating of 1800 W and operates at a voltage of 240 V. What is the resistance of the motor?
a) 2 ohms
b) 3.2 ohms
c) 4.8 ohms
d) 6.4 ohms

Answer: d) 6.4 ohms
Explanation: First, we need to find the current using the power formula P = IV. In this case, I = 1800 W / 240 V = 7.5 A. Then, we can use Ohm's Law (R = V / I) to calculate the resistance. R = 240 V / 7.5 A = 6.4 ohms.

291. What is the primary factor contributing to voltage drop in an electrical circuit?
a) Current
b) Resistance
c) Power
d) Frequency

Answer: b) Resistance
Explanation: Voltage drop is primarily caused by the resistance of conductors in an electrical circuit. As current flows through the conductors, some energy is lost as heat, resulting in a decrease in voltage across the circuit.

292. Why is it essential to consider voltage drop when designing electrical installations?
a) To ensure adequate voltage for all devices
b) To reduce energy consumption
c) To comply with local electrical codes
d) All of the above

Answer: d) All of the above
Explanation: Considering voltage drop is essential to ensure that all devices in the installation receive adequate voltage for proper operation, to reduce energy consumption (and related costs), and to comply with local electrical codes that set limits on allowable voltage drop.

293. In a single-phase circuit with a 120 V source, a resistance of 0.5 ohms, and a current of 10 A, what is the voltage drop across the circuit?
a) 5 V
b) 10 V
c) 20 V
d) 50 V

Answer: a) 5 V
Explanation: To calculate the voltage drop, use Ohm's Law (V = IR). In this case, the voltage drop is V = 10 A × 0.5 ohms = 5 V.

294. How can the voltage drop in a circuit be minimized?
a) By using smaller conductors
b) By increasing the resistance of the circuit
c) By using larger conductors
d) By increasing the voltage of the source

Answer: c) By using larger conductors
Explanation: Voltage drop can be minimized by using larger conductors, which have less resistance and thus cause less energy loss as heat. This will help ensure adequate voltage for all devices in the installation.

295. What is the maximum allowable voltage drop in a branch circuit, according to the National Electrical Code (NEC)?
a) 2%
b) 3%
c) 5%
d) 10%

Answer: b) 3%
Explanation: The National Electrical Code (NEC) recommends a maximum voltage drop of 3% for branch circuits. This limit ensures that devices receive adequate voltage for proper operation and helps maintain energy efficiency in the electrical system.

296. What is the primary factor used to determine the size of a conductor for a specific electrical installation?
a) Voltage
b) Resistance
c) Ampacity
d) Inductance

Answer: c) Ampacity
Explanation: The ampacity, or current-carrying capacity, of a conductor is the primary factor used to determine its size. A conductor must have sufficient ampacity to safely carry the maximum current expected in the circuit without overheating.

297. How does temperature affect the ampacity of a conductor?
a) Higher temperatures increase ampacity
b) Higher temperatures decrease ampacity
c) Temperature has no effect on ampacity
d) The relationship between temperature and ampacity is unpredictable

Answer: b) Higher temperatures decrease ampacity
Explanation: Higher temperatures decrease the ampacity of a conductor due to increased resistance, which can cause overheating and potential failure. Therefore, it's essential to consider the operating temperature when sizing conductors.

298. According to the NEC, what is the minimum ampacity for a conductor that will carry a continuous load of 20 A in an installation with an ambient temperature of 30°C and THHN insulation?
a) 20 A
b) 24 A
c) 25 A
d) 30 A

Answer: c) 25 A
Explanation: For continuous loads, the NEC requires conductors to have an ampacity of at least 125% of the load. In this case, 20 A × 1.25 = 25 A. With an ambient temperature of 30°C and THHN insulation, a conductor with a minimum ampacity of 25 A is required.

299. What effect does conductor insulation have on its ampacity?
a) Insulation type does not affect ampacity
b) Thicker insulation increases ampacity
c) Different insulation materials have different temperature ratings, affecting ampacity
d) Only the color of the insulation affects ampacity

Answer: c) Different insulation materials have different temperature ratings, affecting ampacity
Explanation: Different insulation materials have different temperature ratings, which affect a conductor's ampacity. A higher temperature rating allows a conductor to safely carry more current without overheating.

300. If a circuit has a maximum load of 40 A and requires a conductor with a 75°C insulation rating, which conductor size would be most appropriate according to the NEC?
a) 8 AWG
b) 6 AWG
c) 4 AWG
d) 2 AWG

Answer: b) 6 AWG
Explanation: According to the NEC, a 6 AWG copper conductor with a 75°C insulation rating has an ampacity of 65 A, which is sufficient to carry the 40 A maximum load. An 8 AWG conductor would have an ampacity of 50 A, which may not be enough in some cases, while the larger 4 AWG and 2 AWG conductors would exceed the necessary ampacity and may not be cost-effective.

301. What is the primary purpose of overcurrent protection devices in electrical circuits?
a) To increase circuit efficiency
b) To protect conductors from overheating and potential damage
c) To regulate voltage fluctuations
d) To reduce electrical noise

Answer: b) To protect conductors from overheating and potential damage
Explanation: Overcurrent protection devices, such as fuses and circuit breakers, are designed to protect conductors from overheating and potential damage by interrupting the flow of current in the event of an overcurrent condition, such as a short circuit or overload.

302. How is the required rating for an overcurrent protection device determined?
a) By the maximum voltage in the circuit
b) By the minimum voltage in the circuit
c) By the conductor's ampacity
d) By the circuit's inductive reactance

Answer: c) By the conductor's ampacity
Explanation: The required rating for an overcurrent protection device is determined by the conductor's ampacity, ensuring that the device will interrupt the current flow before the conductor becomes dangerously overheated.

303. Which of the following types of overcurrent protection devices can be reset after a fault has been cleared?
a) Fuses
b) Circuit breakers
c) Thermal relays
d) Surge protectors

Answer: b) Circuit breakers
Explanation: Circuit breakers can be reset after a fault has been cleared, unlike fuses, which must be replaced. This makes circuit breakers more convenient for protecting circuits, as they can be easily reset after the fault condition has been resolved.

304. A circuit has a continuous load of 18 A and uses a conductor with an ampacity of 20 A. What is the appropriate size for an overcurrent protection device in this scenario, according to the NEC?
a) 15 A
b) 20 A
c) 25 A
d) 30 A

Answer: b) 20 A
Explanation: According to the NEC, the overcurrent protection device should not exceed the conductor's ampacity. In this case, a 20 A device would be appropriate, as it matches the conductor's ampacity without exceeding it.

305. In a motor circuit with a full-load current of 10 A and a 1.25 service factor, what is the minimum size of the overcurrent protection device, assuming a dual-element time-delay fuse is used?
a) 15 A
b) 20 A
c) 25 A
d) 30 A

Answer: c) 25 A. Explanation: To determine the minimum size of the overcurrent protection device, multiply the motor's full-load current by the service factor and the NEC allowance for dual-element time-delay fuses (175%). In this case, 10 A × 1.25 × 1.75 = 21.875 A. The closest standard size that meets this requirement is a 25 A fuse.

306. What is a short circuit in an electrical system?
a) A decrease in resistance, leading to an increase in current flow
b) A sudden increase in resistance, leading to a decrease in current flow
c) An unintentional connection between two points of differing potential
d) A normal operating condition with balanced current flow

Answer: c) An unintentional connection between two points of differing potential
Explanation: A short circuit is an unintended connection between two points of differing potential in an electrical system. This results in a low-resistance path for the current to flow, causing a sudden increase in current, which can lead to damage or even fire.

307. What is the primary factor influencing the magnitude of a short circuit current?
a) Circuit impedance
b) Circuit resistance
c) Circuit inductance
d) Circuit capacitance

Answer: a) Circuit impedance
Explanation: The magnitude of a short circuit current is primarily influenced by the circuit impedance, which includes both resistance and reactance components. Lower impedance results in higher short circuit currents, while higher impedance leads to lower short circuit currents.

308. In a three-phase electrical system, what is the most common type of short circuit?
a) Line-to-line
b) Line-to-ground
c) Double line-to-ground
d) Three-phase

Answer: b) Line-to-ground
Explanation: In a three-phase electrical system, the most common type of short circuit is a line-to-ground fault. This occurs when one of the phase conductors unintentionally comes into contact with the ground or a grounded object, creating an unintended low-resistance path for the current.

309. If a single-phase system has a short circuit current of 5,000 A and an impedance of 0.1 ohms, what is the system voltage?
a) 500 V
b) 1,000 V
c) 1,500 V
d) 2,000 V

Answer: a) 500 V
Explanation: To calculate the system voltage, use Ohm's Law ($V = I \times Z$), where V is the voltage, I is the short circuit current, and Z is the impedance. In this case, V = 5,000 A × 0.1 ohms = 500 V.

310. What is the main purpose of calculating short circuit currents in electrical systems?
a) To ensure proper sizing of conductors
b) To maintain voltage stability
c) To properly size and select overcurrent protection devices
d) To calculate energy efficiency

Answer: c) To properly size and select overcurrent protection devices
Explanation: Calculating short circuit currents in electrical systems is crucial for properly sizing and selecting overcurrent protection devices, such as fuses and circuit breakers. These devices must be capable of interrupting the maximum short circuit current that can occur in the system to protect equipment and personnel from potential harm.

311. What is the primary purpose of calculating electrical loads for residential and commercial installations?
a) To ensure voltage stability
b) To determine the appropriate size of electrical service equipment
c) To calculate energy consumption for billing purposes
d) To design energy-efficient lighting systems

Answer: b) To determine the appropriate size of electrical service equipment
Explanation: Calculating electrical loads for residential and commercial installations is essential for determining the appropriate size of electrical service equipment, such as transformers, panels, and conductors. Properly sized equipment ensures safe and reliable operation while avoiding overloading and potential hazards.

312. What is a demand factor in the context of electrical load calculations?
a) The ratio of the maximum demand to the total connected load
b) The percentage of the total load that is operating simultaneously
c) The ratio of the total connected load to the maximum demand
d) The percentage of the maximum demand that is operating simultaneously

Answer: a) The ratio of the maximum demand to the total connected load
Explanation: The demand factor is the ratio of the maximum demand to the total connected load. It accounts for the fact that not all electrical devices will be operating at full capacity or simultaneously, allowing for a more accurate estimation of the actual load on the electrical system.

313. What is a diversity factor in electrical load calculations?
a) The ratio of the sum of individual maximum loads to the overall maximum load
b) The ratio of the overall maximum load to the sum of individual maximum loads
c) The percentage of the total load that is operating simultaneously
d) The percentage of the maximum demand that is operating simultaneously

Answer: b) The ratio of the overall maximum load to the sum of individual maximum loads
Explanation: The diversity factor is the ratio of the overall maximum load to the sum of individual maximum loads in an electrical system. It accounts for the fact that different loads have varying operating patterns, and it is unlikely that they will all operate at their maximum capacity simultaneously.

314. In a residential installation, if the total connected load is 20,000 VA and the demand factor is 0.7, what is the estimated maximum demand?
a) 6,000 VA
b) 14,000 VA
c) 20,000 VA
d) 28,571 VA

Answer: b) 14,000 VA
Explanation: To calculate the estimated maximum demand, multiply the total connected load by the demand factor. In this case, 20,000 VA × 0.7 = 14,000 VA.

315. If a commercial installation has a total connected load of 50,000 VA, a demand factor of 0.6, and a diversity factor of 1.25, what is the estimated maximum demand?
a) 30,000 VA
b) 37,500 VA
c) 40,000 VA
d) 62,500 VA

Answer: c) 40,000 VA
Explanation: First, calculate the estimated maximum demand using the demand factor: 50,000 VA × 0.6 = 30,000 VA. Next, apply the diversity factor to the estimated maximum demand: 30,000 VA × 1.25 = 40,000 VA.

316. What is the primary function of a transformer in an electrical system?
a) To regulate voltage fluctuations
b) To convert AC voltage to DC voltage
c) To step up or step down voltage levels
d) To provide electrical isolation between circuits

Answer: c) To step up or step down voltage levels
Explanation: The primary function of a transformer is to step up or step down voltage levels in an electrical system. This is achieved by utilizing the principle of electromagnetic induction between two or more windings.

317. When selecting a transformer for an application, which of the following factors should be considered?
a) Voltage rating
b) Power rating
c) Temperature rating
d) All of the above

Answer: d) All of the above
Explanation: When selecting a transformer for an application, factors such as voltage rating, power rating, and temperature rating must be considered to ensure proper operation and safety.

318. Using the transformer turns ratio, how can the secondary voltage be calculated?
a) Primary voltage × (primary turns / secondary turns)
b) Primary voltage × (secondary turns / primary turns)
c) Primary voltage / (primary turns / secondary turns)
d) Primary voltage / (secondary turns / primary turns)

Answer: b) Primary voltage × (secondary turns / primary turns)
Explanation: To calculate the secondary voltage using the transformer turns ratio, multiply the primary voltage by the ratio of secondary turns to primary turns.

319. In a step-down transformer, if the primary voltage is 480V, and the transformer has a turns ratio of 4:1, what is the secondary voltage?
a) 120V
b) 240V
c) 480V
d) 1920V

Answer: a) 120V
Explanation: To calculate the secondary voltage, use the formula: Primary voltage × (secondary turns / primary turns). In this case, 480V × (1 / 4) = 120V.

320. If a transformer has a primary current of 15A, a primary voltage of 240V, and a secondary voltage of 480V, what is the secondary current?
a) 7.5A
b) 15A
c) 30A
d) 60A

Answer: a) 7.5A
Explanation: To calculate the secondary current, use the formula: Primary current × (primary voltage / secondary voltage). In this case, 15A × (240V / 480V) = 7.5A.

321. When selecting an electric motor for an application, which of the following factors should be considered?
a) Horsepower
b) Efficiency
c) Motor type and speed
d) All of the above

Answer: d) All of the above
Explanation: When selecting an electric motor for an application, factors such as horsepower, efficiency, motor type, and speed must be considered to ensure proper operation and match the requirements of the specific application.

322. What is the relationship between horsepower, torque, and motor speed?
a) Horsepower = Torque × Speed
b) Horsepower = Torque × Speed / 5252
c) Horsepower = Torque / Speed × 5252
d) Horsepower = Torque / (Speed × 5252)

Answer: b) Horsepower = Torque × Speed / 5252
Explanation: The relationship between horsepower, torque, and motor speed is given by the formula: Horsepower = Torque × Speed / 5252, where torque is in pound-feet and speed is in RPM.

323. If a motor has a power rating of 10 HP and operates at an efficiency of 90%, what is the input power required?
a) 9 kW
b) 10 kW
c) 11.1 kW
d) 12 kW

Answer: c) 11.1 kW
Explanation: To calculate the input power required, divide the motor power rating by the efficiency. In this case, (10 HP × 0.746 kW/HP) / 0.9 = 11.1 kW.

324. A motor has a full-load current of 25A, a voltage rating of 480V, and a power factor of 0.8. What is the motor's apparent power?
a) 6 kVA
b) 7.5 kVA
c) 12 kVA
d) 15 kVA

Answer: d) 15 kVA. Explanation: To calculate the motor's apparent power, use the formula: Apparent power = Voltage × Current. In this case, 480V × 25A = 12,000 VA, or 12 kVA.

325. A 3-phase motor has a voltage rating of 240V, a power rating of 15 HP, and an efficiency of 90%. What is the motor's full-load current?
a) 28.4 A
b) 37.8 A
c) 45.2 A
d) 56.8 A

Answer: b) 37.8 A. Explanation: To calculate the motor's full-load current, use the formula: Full-load current = (Power × 1000) / (√3 × Voltage × Efficiency × Power Factor). Assume a power factor of 0.85. In this case, (15 HP × 0.746 kW/HP × 1000) / (√3 × 240V × 0.9 × 0.85) ≈ 37.8 A.

326. When calculating illuminance levels for a room, which of the following factors should be considered?
a) Room dimensions
b) Reflectance of surfaces
c) Required lighting level
d) All of the above

Answer: d) All of the above. Explanation: When calculating illuminance levels for a room, factors such as room dimensions, reflectance of surfaces, and required lighting levels must be considered to ensure appropriate lighting for the specific application.

327. Which of the following formulas is used to calculate the number of luminaires required for a space?
a) Number of luminaires = (Illuminance × Area) / (Luminaire output × Utilization factor × Maintenance factor)
b) Number of luminaires = (Illuminance × Area) × (Luminaire output × Utilization factor × Maintenance factor)
c) Number of luminaires = (Illuminance × Luminaire output) / (Area × Utilization factor × Maintenance factor)
d) Number of luminaires = (Illuminance × Utilization factor × Maintenance factor) / (Area × Luminaire output)

Answer: a) Number of luminaires = (Illuminance × Area) / (Luminaire output × Utilization factor × Maintenance factor)
Explanation: The number of luminaires required for a space is calculated using the formula: Number of luminaires = (Illuminance × Area) / (Luminaire output × Utilization factor × Maintenance factor).

328. What is the purpose of the maintenance factor in lighting calculations?
a) To account for the decrease in luminaire output over time
b) To account for dirt accumulation on the luminaire surface
c) To account for changes in room reflectances over time
d) All of the above

Answer: d) All of the above
Explanation: The maintenance factor accounts for the decrease in luminaire output over time, dirt accumulation on the luminaire surface, and changes in room reflectances over time, ensuring proper lighting levels are maintained throughout the life of the lighting system.

329. A room requires an illuminance level of 500 lux. The room dimensions are 10m x 5m, and the selected luminaire has an output of 4000 lumens. If the utilization factor is 0.6 and the maintenance factor is 0.8, how many luminaires are required?
a) 5
b) 7
c) 10
d) 12

Answer: c) 10
Explanation: Using the formula for the number of luminaires, (500 lux × 50m$^2$) / (4000 lumens × 0.6 × 0.8) ≈ 10.42. Since you cannot have a fraction of a luminaire, round up to the nearest whole number, which is 10.

330. If a lighting system consumes 2 kW and operates for 10 hours per day, what is the daily energy consumption?
a) 20 kWh
b) 200 kWh
c) 2,000 kWh
d) 20,000 kWh

Answer: a) 20 kWh
Explanation: The daily energy consumption can be calculated by multiplying the power consumption by the operating hours. In this case, 2 kW × 10 hours = 20 kWh.

331. Which of the following factors can impact the energy efficiency of an electrical installation?
a) Equipment efficiency
b) System design
c) Operational practices
d) All of the above

Answer: d) All of the above
Explanation: The energy efficiency of an electrical installation can be impacted by various factors, including equipment efficiency, system design, and operational practices. By considering all of these factors, engineers can design more energy-efficient systems.

332. How is energy consumption typically calculated for a given electrical device?
a) Power rating × Operating hours
b) Power rating × Operating hours × Efficiency
c) Voltage × Current × Operating hours
d) Voltage × Current × Operating hours × Efficiency

Answer: a) Power rating × Operating hours
Explanation: Energy consumption for a given electrical device is typically calculated by multiplying its power rating (in watts or kilowatts) by the number of operating hours.

333. What is the primary purpose of a power factor correction in an electrical system?
a) To reduce energy consumption
b) To improve voltage stability
c) To minimize harmonics
d) To reduce equipment wear

Answer: a) To reduce energy consumption
Explanation: The primary purpose of power factor correction is to reduce energy consumption by minimizing the reactive power component in an electrical system. This improves the overall efficiency of the system and reduces energy costs.

334. In an electrical system, what is the primary advantage of using high-efficiency motors?
a) Lower energy consumption
b) Increased torque
c) Longer lifespan
d) Faster startup times

Answer: a) Lower energy consumption
Explanation: High-efficiency motors consume less energy compared to standard motors for the same amount of mechanical work, resulting in reduced energy costs and lower environmental impact.

335. An electrical device with a power rating of 1.5 kW operates for 8 hours per day. If the electricity cost is $0.12 per kWh, what is the monthly energy cost for this device?
a) $43.20
b) $57.60
c) $72.00
d) $86.40

Answer: a) $43.20
Explanation: First, calculate the daily energy consumption: 1.5 kW × 8 hours = 12 kWh. Then, calculate the daily cost: 12 kWh × $0.12 per kWh = $1.44. Finally, calculate the monthly cost (assuming 30 days in a month): $1.44 × 30 = $43.20.

336. An electrical circuit consists of a 24 V power supply and a 12 Ω resistor. What is the current flowing through the circuit?
a) 0.5 A
b) 1 A
c) 2 A
d) 4 A

Answer: c) 2 A
Explanation: Using Ohm's Law (V = IR), we can find the current: I = V / R = 24 V / 12 Ω = 2 A.

337. A single-phase motor has an efficiency of 85% and a power factor of 0.9. If the motor consumes 3 kW of power, what is the apparent power (in kVA) drawn by the motor?
a) 3.529 kVA
b) 3.922 kVA
c) 4.118 kVA
d) 4.706 kVA

Answer: a) 3.529 kVA
Explanation: To find the apparent power, first find the real power: P_real = P_consumed / efficiency = 3 kW / 0.85 = 3.529 kW. Then, find the apparent power: P_apparent = P_real / power factor = 3.529 kW / 0.9 = 3.921 kVA (rounded to three decimal places).

338. A lighting system in a commercial building consists of 20 fixtures, each with a power rating of 60 W. If the system operates for 10 hours per day, what is the daily energy consumption in kWh?
a) 1.2 kWh
b) 6 kWh
c) 12 kWh
d) 60 kWh

Answer: c) 12 kWh
Explanation: First, find the total power consumption: 20 fixtures × 60 W = 1,200 W (or 1.2 kW). Then, calculate the daily energy consumption: 1.2 kW × 10 hours = 12 kWh.

339. A transformer has a primary voltage of 480 V, a secondary voltage of 240 V, and a primary current of 10 A. What is the secondary current?
a) 5 A
b) 10 A
c) 20 A
d) 40 A

Answer: c) 20 A
Explanation: Using the transformer turns ratio formula, we can find the secondary current: V_primary / V_secondary = I_secondary / I_primary. Solving for I_secondary: I_secondary = I_primary × (V_primary / V_secondary) = 10 A × (480 V / 240 V) = 20 A.

340. A three-phase, 480 V motor has a power factor of 0.8 and consumes 40 kW of power. What is the total current drawn by the motor?
a) 48.08 A
b) 60.10 A
c) 72.12 A
d) 80.16 A

Answer: b) 60.10 A
Explanation: First, find the apparent power: P_apparent = P_real / power factor = 40 kW / 0.8 = 50 kVA. Then, calculate the total current using the formula I_total = P_apparent × 1,000 / (√3 × V), where V is the line voltage: I_total = 50,000 VA / (√3 × 480 V) ≈ 60.10 A (rounded to two decimal places).

341. What is the minimum working space required in front of an electrical panel with a voltage between 150 V and 600 V, according to the National Electrical Code (NEC)?
a) 24 inches
b) 30 inches
c) 36 inches
d) 42 inches

Answer: c) 36 inches
Explanation: The NEC requires a minimum working space of 36 inches in front of electrical panels with a voltage between 150 V and 600 V for safe operation and maintenance.

342. When selecting an electrical panel size, what factor should be considered?
a) The number of circuits in the system
b) The color of the panel
c) The material of the panel
d) The distance from the main power source

Answer: a) The number of circuits in the system
Explanation: The size of the electrical panel should be based on the number of circuits in the system. A larger panel is needed to accommodate more circuits and provide room for future expansion.

343. Which of the following locations is NOT suitable for installing an electrical panel?
a) In a well-ventilated utility room
b) In a dry, accessible basement
c) In a bathroom
d) In a garage with proper clearance

Answer: c) In a bathroom
Explanation: Electrical panels should not be installed in bathrooms due to the potential hazards associated with water and moisture, as well as the need for a dedicated, non-humid workspace.

344. When selecting the location of an electrical panel, which of the following factors should be considered?
a) Accessibility for maintenance and operation
b) Proximity to water sources
c) Distance from windows and doors
d) All of the above

Answer: a) Accessibility for maintenance and operation
Explanation: The location of an electrical panel should provide easy access for maintenance and operation. While factors like proximity to water sources and distance from windows and doors can be important for safety, the primary consideration should be accessibility.

345. What is the main purpose of installing a main breaker in an electrical panel?
a) To provide an additional layer of protection
b) To disconnect power to the entire panel
c) To monitor the power usage of the building
d) To distribute power evenly among the circuits

Answer: b) To disconnect power to the entire panel
Explanation: The main breaker in an electrical panel serves as a means to disconnect power to the entire panel. This allows for safe maintenance, troubleshooting, or emergency shutdown of the entire electrical system.

346. Which type of conduit is most commonly used for indoor commercial and industrial installations due to its durability and flexibility?
a) Rigid Metal Conduit (RMC)
b) Electrical Metallic Tubing (EMT)
c) Flexible Metal Conduit (FMC)
d) Polyvinyl Chloride (PVC) Conduit

Answer: b) Electrical Metallic Tubing (EMT)
Explanation: EMT is a popular choice for indoor commercial and industrial installations because it is durable, lightweight, and easier to install than RMC. It also provides a good balance between flexibility and protection for electrical conductors.

347. What is the main purpose of using a conduit in electrical installations?
a) To reduce electromagnetic interference
b) To protect conductors from damage
c) To reduce the risk of short circuits
d) To improve the aesthetics of the installation

Answer: b) To protect conductors from damage
Explanation: The primary purpose of using a conduit in electrical installations is to protect conductors from physical damage, such as impact or abrasion. Conduits also help to organize and route conductors more effectively.

348. When selecting a cable type for a given electrical installation, which factor should be considered?
a) The size of the building
b) The type of load being served
c) The color of the cable insulation
d) The distance from the main electrical panel

Answer: b) The type of load being served
Explanation: The type of load being served should be considered when selecting a cable type. Different cable types are designed to handle different types of loads, such as lighting, appliances, or motors, and have different insulation and conductor materials to suit the application.

349. Which of the following is a common type of wire connection used in electrical installations?
a) Twisted pair
b) Solder joint
c) Wire nut
d) Bolted joint

Answer: c) Wire nut
Explanation: Wire nuts are a common type of wire connection used in electrical installations. They are used to securely join two or more conductors together and provide insulation to protect the connection from exposure to air and moisture.

350. Which of the following is NOT a recommended practice when making wire connections in an electrical installation?
a) Ensuring that all conductors are securely fastened
b) Using a wire connector that is appropriately sized for the conductors
c) Leaving exposed copper wire outside the wire connector
d) Tightly twisting the conductors together before applying a wire connector

Answer: c) Leaving exposed copper wire outside the wire connector
Explanation: Exposed copper wire outside the wire connector can create a potential hazard, such as a short circuit or electrical shock. When making wire connections, it is important to ensure that all conductors are securely fastened and fully enclosed within the wire connector.

351. What is the primary purpose of grounding in an electrical installation?
a) To provide a path for lightning to follow
b) To reduce the risk of electric shock
c) To improve energy efficiency
d) To reduce electromagnetic interference

Answer: b) To reduce the risk of electric shock
Explanation: Grounding is crucial for providing a low-resistance path for fault current to flow to earth, which helps to reduce the risk of electric shock. This is important for the safety of people and equipment in the event of a fault or malfunction in the electrical system.

352. Which of the following is a suitable grounding electrode for an electrical installation?
a) A wooden pole
b) A metal water pipe
c) An insulated copper wire
d) A concrete-encased electrode

Answer: d) A concrete-encased electrode
Explanation: A concrete-encased electrode, also known as a "Ufer" ground, is a suitable grounding electrode for an electrical installation. It is embedded in the building's concrete foundation and provides a low-resistance path for fault current to flow to the earth.

353. What is the primary function of bonding in an electrical installation?
a) To ensure all metallic parts are at the same electrical potential
b) To create a path for lightning to follow
c) To increase the efficiency of the electrical system
d) To prevent electrical fires

Answer: a) To ensure all metallic parts are at the same electrical potential
Explanation: Bonding ensures that all metallic parts of the electrical system are at the same electrical potential. This helps to minimize the risk of electric shock, arcing, and other potential hazards by providing a low-resistance path for fault current to flow.

354. Which type of conductor is typically used for grounding in an electrical installation?
a) Aluminum
b) Copper
c) Steel
d) Silver

Answer: b) Copper
Explanation: Copper is typically used as the grounding conductor in electrical installations because it offers high conductivity, low resistance, and is resistant to corrosion. This ensures a reliable and effective grounding system.

355. In an electrical installation, what is the purpose of connecting the neutral conductor to the ground at the service entrance?
a) To protect against lightning strikes
b) To reduce the risk of electric shock
c) To provide a return path for fault current
d) To stabilize the voltage on the system

Answer: c) To provide a return path for fault current
Explanation: Connecting the neutral conductor to the ground at the service entrance, also known as the main bonding jumper, provides a return path for fault current. This helps to clear faults quickly and safely, minimizing the risk of damage to equipment and the risk of electric shock.

356. What is the first step that should be taken before installing an electrical device such as a receptacle, switch, or light fixture?
a) Turn off the power at the main panel
b) Gather all necessary tools and materials
c) Remove the old device, if applicable
d) Test the device to ensure proper operation

Answer: a) Turn off the power at the main panel
Explanation: The first and most important step before installing any electrical device is to turn off the power at the main panel. This ensures the safety of the installer and prevents the risk of electric shock or other accidents.

357. Which type of receptacle should be installed in areas where water is likely to be present, such as bathrooms and kitchens?
a) Standard receptacle
b) GFCI (Ground Fault Circuit Interrupter) receptacle
c) AFCI (Arc Fault Circuit Interrupter) receptacle
d) Tamper-resistant receptacle

Answer: b) GFCI (Ground Fault Circuit Interrupter) receptacle
Explanation: GFCI receptacles are designed to protect against electrical shock in areas where water is likely to be present. They detect imbalances in current flow and quickly disconnect power if a ground fault is detected.

358. When installing a switch, which wire should be connected to the screw terminal marked "common"?
a) Ground wire
b) Neutral wire
c) Hot wire
d) Traveler wire

Answer: c) Hot wire
Explanation: When installing a switch, the hot wire (typically black or red) should be connected to the screw terminal marked "common." This terminal is responsible for providing power to the switch and ensuring the proper function of the device.

359. What is the primary purpose of using a box extender when installing an electrical device?
a) To increase the capacity of the electrical box
b) To provide a more secure mounting for the device
c) To ensure proper grounding
d) To bring the electrical box flush with the wall surface

Answer: d) To bring the electrical box flush with the wall surface
Explanation: A box extender is used to bring the electrical box flush with the wall surface when installing an electrical device. This ensures proper alignment of the device with the wall and provides a secure and professional installation.

360. When installing a light fixture, which wire should be connected to the fixture's grounding screw or green wire?
a) Hot wire
b) Neutral wire
c) Ground wire
d) Load wire

Answer: c) Ground wire
Explanation: When installing a light fixture, the ground wire (typically green or bare copper) should be connected to the fixture's grounding screw or green wire. This provides a safe path for fault current and helps prevent the risk of electric shock.

361. When selecting a circuit breaker for an electrical system, which of the following factors must be considered?
a) The type of load being protected
b) The voltage rating of the circuit
c) The amperage rating of the circuit
d) All of the above

Answer: d) All of the above
Explanation: When selecting a circuit breaker, it is important to consider the type of load being protected, the voltage rating of the circuit, and the amperage rating of the circuit. These factors ensure that the circuit breaker is suitable for the application and provides the necessary protection.

362. Which type of fuse is designed to prevent the overcurrent protection from opening during a short circuit?
a) Fast-acting fuse
b) Time-delay fuse
c) Dual-element fuse
d) High-interrupting fuse

Answer: b) Time-delay fuse
Explanation: Time-delay fuses are designed to prevent the overcurrent protection from opening during a short circuit. They have a built-in delay that allows them to handle temporary overcurrent conditions, such as motor startup, without opening the circuit.

363. What is the purpose of an interruption rating on a circuit breaker?
a) To indicate the maximum current the breaker can handle without tripping
b) To indicate the minimum current that will cause the breaker to trip
c) To indicate the maximum voltage the breaker can handle
d) To indicate the maximum short-circuit current the breaker can safely interrupt

Answer: d) To indicate the maximum short-circuit current the breaker can safely interrupt
Explanation: The interruption rating on a circuit breaker indicates the maximum short-circuit current the breaker can safely interrupt. It is important to ensure that the interruption rating is sufficient for the electrical system in which the breaker is installed to avoid potential damage or hazards.

364. When installing a circuit breaker, what should you do if the breaker does not fit securely in the panel?
a) Force the breaker into place
b) Use a different type of breaker that fits
c) Modify the panel to accommodate the breaker
d) Consult the panel manufacturer for compatibility

Answer: d) Consult the panel manufacturer for compatibility
Explanation: If a circuit breaker does not fit securely in the panel, you should consult the panel manufacturer for compatibility information. Forcing the breaker into place or modifying the panel could compromise the safety and functionality of the electrical system.

365. In a residential application, which of the following circuit breaker types is most commonly used to protect against both ground faults and arc faults?
a) Standard circuit breaker
b) GFCI (Ground Fault Circuit Interrupter) breaker
c) AFCI (Arc Fault Circuit Interrupter) breaker
d) Dual-function circuit breaker

Answer: d) Dual-function circuit breaker
Explanation: Dual-function circuit breakers are designed to protect against both ground faults and arc faults in residential applications. These breakers combine the functionality of GFCI and AFCI breakers, providing comprehensive protection for the electrical system.

366. Which of the following is an essential first step when troubleshooting an electrical circuit?
a) Replacing the circuit breaker
b) Measuring the voltage across the load
c) Identifying the problem and gathering necessary tools
d) Turning off the power and verifying it is off

Answer: d) Turning off the power and verifying it is off
Explanation: When troubleshooting an electrical circuit, safety should always be the priority. The first step is to turn off the power to the circuit and verify that it is off using a voltage tester or multimeter.

367. What is the primary function of a clamp meter in troubleshooting electrical circuits?
a) To measure voltage
b) To measure resistance
c) To measure current without breaking the circuit
d) To measure capacitance

Answer: c) To measure current without breaking the circuit
Explanation: A clamp meter is a versatile test instrument that allows you to measure current without breaking the circuit. By clamping the meter around a conductor, you can safely and quickly measure the current flowing through it.

368. When measuring resistance in an electrical circuit, what should be done before using a multimeter?
a) Apply power to the circuit
b) Remove power from the circuit
c) Check for continuity
d) Adjust the multimeter's range

Answer: b) Remove power from the circuit
Explanation: When measuring resistance in an electrical circuit, it is crucial to remove power from the circuit before using a multimeter. This ensures safety and prevents potential damage to the multimeter or the circuit being tested.

369. Which of the following is a common cause of an open circuit?
a) A shorted component
b) A blown fuse or tripped circuit breaker
c) A loose or corroded connection
d) Excessive current draw

Answer: c) A loose or corroded connection
Explanation: An open circuit is a break in the continuity of the circuit, preventing current from flowing. A common cause of an open circuit is a loose or corroded connection, which can be identified and repaired during troubleshooting.

370. When troubleshooting a series circuit with multiple loads, which of the following is the best approach to identify a faulty load?
a) Measure the total current in the circuit
b) Measure the voltage across each load
c) Measure the resistance of each load
d) Bypass each load one at a time

Answer: b) Measure the voltage across each load
Explanation: When troubleshooting a series circuit with multiple loads, the best approach to identify a faulty load is to measure the voltage across each load. In a series circuit, the current remains the same through all loads, but the voltage drops across each load will vary. By measuring the voltage drops and comparing them to the expected values, you can identify any load that is not functioning correctly.

371. What is a common cause of an open circuit in a residential electrical installation?
a) A shorted component
b) A tripped GFCI outlet
c) A loose wire connection
d) A grounded conductor

Answer: c) A loose wire connection
Explanation: A loose wire connection is a common cause of an open circuit in residential electrical installations. The loose connection interrupts the continuity of the circuit, preventing current flow.

372. What is the most likely cause of flickering lights in a commercial building?
a) Overloaded circuits
b) Voltage fluctuations
c) Faulty light fixtures
d) Ground faults

Answer: b) Voltage fluctuations
Explanation: Voltage fluctuations are the most likely cause of flickering lights in a commercial building. These fluctuations can be caused by a variety of factors, including large loads starting and stopping, power supply issues, or loose connections within the electrical system.

373. Which of the following is the best method for identifying the cause of a short circuit?
a) Visual inspection of the circuit
b) Measuring voltage across the circuit
c) Measuring current in the circuit
d) Continuity testing of the circuit components

Answer: d) Continuity testing of the circuit components
Explanation: Continuity testing is the best method for identifying the cause of a short circuit. By testing the circuit components for unintended continuity, you can locate the short circuit and repair the issue.

374. What is the most appropriate solution for an overloaded circuit?
a) Replace the circuit breaker with a higher-rated one
b) Remove some load from the circuit
c) Tighten all connections within the circuit
d) Install additional grounding electrodes

Answer: b) Remove some load from the circuit
Explanation: The most appropriate solution for an overloaded circuit is to remove some load from the circuit, thus preventing excessive current flow and the potential for overheating or fire hazards.

375. A GFCI outlet keeps tripping in a residential installation. What is the most likely cause?
a) An open circuit
b) A short circuit
c) A ground fault
d) A voltage fluctuation

Answer: c) A ground fault
Explanation: A GFCI outlet is designed to trip when it detects a ground fault, which is an unintended connection between a live conductor and a grounded surface. If a GFCI outlet keeps tripping, it is most likely detecting a ground fault in the circuit.

376. When working on electrical installations, which of the following is the primary reason for wearing personal protective equipment (PPE)?
a) To comply with OSHA regulations
b) To protect against electrical shock
c) To prevent damage to equipment
d) To avoid liability in case of an accident

Answer: b) To protect against electrical shock
Explanation: The primary reason for wearing PPE when working on electrical installations is to protect against electrical shock. PPE helps to minimize the risk of injury and ensures that workers are protected from hazards associated with electrical work.

377. Which of the following is an essential safety practice when working on energized electrical equipment?
a) Work with one hand while keeping the other hand in your pocket
b) Use metal tools to ensure proper grounding
c) Wear jewelry to enhance electrical conductivity
d) Always assume that the equipment is de-energized

Answer: a) Work with one hand while keeping the other hand in your pocket
Explanation: Working with one hand while keeping the other hand in your pocket is an essential safety practice when working on energized electrical equipment. This reduces the risk of creating a path for electrical current to flow through the body, potentially causing injury or death.

378. Which of the following should be used to ensure electrical safety when working on or near live conductors?
a) Non-contact voltage tester
b) Standard multimeter
c) Voltage tick tester
d) Insulation resistance tester

Answer: a) Non-contact voltage tester
Explanation: A non-contact voltage tester is the safest tool for verifying the presence of voltage in live conductors. It does not require physical contact with the conductor, reducing the risk of electrical shock or injury.

379. According to OSHA regulations, which of the following is a requirement for properly maintaining electrical equipment?
a) Frequent visual inspections
b) Annual replacement of all components
c) Continuously monitoring power usage
d) Cleaning with water and soap


Answer: a) Frequent visual inspections
Explanation: OSHA regulations require that electrical equipment be properly maintained, including frequent visual inspections. Regular inspections help identify potential hazards and ensure that equipment is in good working order.

380. What is the purpose of using lockout/tagout procedures during electrical maintenance work?
a) To prevent unauthorized access to the work area
b) To ensure that equipment is properly labeled
c) To prevent the accidental re-energizing of equipment
d) To prevent theft of tools and equipment


Answer: c) To prevent the accidental re-energizing of equipment
Explanation: Lockout/tagout procedures are used during electrical maintenance work to prevent the accidental re-energizing of equipment. These procedures help protect workers from electrical hazards by ensuring that equipment remains de-energized while maintenance is being performed.

381. What is the primary goal of performing preventive maintenance on electrical systems?
a) To minimize equipment downtime
b) To comply with local electrical codes
c) To increase energy efficiency
d) To reduce maintenance costs


Answer: a) To minimize equipment downtime
Explanation: The primary goal of performing preventive maintenance on electrical systems is to minimize equipment downtime. Regular maintenance helps to identify and address potential issues before they cause equipment failures or other problems, ensuring the reliability of the electrical system.

382. Which of the following is a common preventive maintenance task for electrical systems?
a) Tightening loose connections
b) Replacing all components every year
c) Painting electrical panels
d) Washing electrical components with water

Answer: a) Tightening loose connections
Explanation: Tightening loose connections is a common preventive maintenance task for electrical systems. Loose connections can cause excessive heat, arcing, and potentially equipment failure, so regular inspections and tightening are important for maintaining system reliability.

383. Which device should be regularly tested as part of a preventive maintenance program for electrical systems?
a) Circuit breakers
b) Voltage transformers
c) Panelboards
d) Conduit

Answer: a) Circuit breakers
Explanation: Circuit breakers should be regularly tested as part of a preventive maintenance program for electrical systems. Regular testing ensures that circuit breakers are functioning correctly and will trip when needed to protect electrical equipment and personnel.

384. Which of the following is an important aspect of a preventive maintenance program for motors?
a) Regular lubrication
b) Frequent replacement of motor brushes
c) Continuously running the motor at maximum load
d) Ignoring unusual noises or vibrations

Answer: a) Regular lubrication
Explanation: Regular lubrication is an important aspect of a preventive maintenance program for motors. Proper lubrication helps to reduce friction, minimize wear, and extend the life of motor components.

385. What is the purpose of performing infrared thermography as part of a preventive maintenance program for electrical systems?
a) To identify overloaded circuits
b) To measure the efficiency of electrical devices
c) To locate sources of electromagnetic interference
d) To detect hot spots indicating potential problems

Answer: d) To detect hot spots indicating potential problems
Explanation: The purpose of performing infrared thermography as part of a preventive maintenance program for electrical systems is to detect hot spots indicating potential problems. Hot spots can be caused by loose connections, overloaded circuits, or other issues, and infrared thermography can help identify these areas before they lead to equipment failure or other problems.

386. Which of the following symptoms may indicate a failed transformer in an electrical system?
a) Excessive noise from electrical devices
b) Dimming or flickering lights
c) Tripped circuit breakers
d) All of the above

Answer: d) All of the above
Explanation: A failed transformer can cause various issues in an electrical system, including excessive noise from electrical devices, dimming or flickering lights, and tripped circuit breakers. Any of these symptoms may indicate the need to troubleshoot and potentially replace the transformer.

387. When troubleshooting a transformer, what test equipment is commonly used to measure the transformer's primary and secondary voltages?
a) Clamp meter
b) Multimeter
c) Megohmmeter
d) Oscilloscope

Answer: b) Multimeter
Explanation: A multimeter is commonly used to measure the primary and secondary voltages of a transformer when troubleshooting. By measuring these voltages, an electrician can determine if the transformer is functioning properly or if it needs to be replaced.

388. Which of the following factors should be considered when selecting a replacement transformer?
a) Voltage rating
b) KVA rating
c) Physical dimensions
d) All of the above

Answer: d) All of the above. Explanation: When selecting a replacement transformer, it is important to consider the voltage rating, KVA rating, and physical dimensions of the new transformer. These factors ensure that the replacement transformer will meet the requirements of the electrical system and fit within the available space.

389. What is the primary reason to check the insulation resistance of a transformer during troubleshooting?
a) To determine the efficiency of the transformer
b) To identify internal short circuits or winding faults
c) To measure the transformer's current draw
d) To verify the transformer's voltage rating

Answer: b) To identify internal short circuits or winding faults
Explanation: Checking the insulation resistance of a transformer during troubleshooting helps identify internal short circuits or winding faults. These issues can cause the transformer to fail, and identifying them can help determine if the transformer needs to be replaced.

390. When replacing a transformer, which safety precaution should be taken to avoid electrical hazards?
a) Work on the transformer while it is energized
b) Disconnect the transformer from the power source and verify the absence of voltage
c) Use a metal ladder to reach the transformer
d) Use uninsulated tools during the replacement process

Answer: b) Disconnect the transformer from the power source and verify the absence of voltage
Explanation: When replacing a transformer, it is crucial to disconnect the transformer from the power source and verify the absence of voltage. This safety precaution helps to prevent electrical hazards and ensure the safety of the person performing the replacement.

391. Which of the following is a common symptom of a failing electric motor?
a) Excessive vibration
b) Overheating
c) Frequent tripping of circuit breakers or fuses
d) All of the above

Answer: d) All of the above
Explanation: A failing electric motor may exhibit various symptoms, such as excessive vibration, overheating, and frequent tripping of circuit breakers or fuses. Identifying these symptoms can help determine if the motor requires troubleshooting and repair.

392. When troubleshooting an electric motor, which test equipment is commonly used to measure winding resistance and insulation resistance?
a) Multimeter
b) Megohmmeter
c) Clamp meter
d) Oscilloscope

Answer: b) Megohmmeter
Explanation: A megohmmeter is commonly used when troubleshooting an electric motor to measure winding resistance and insulation resistance. These measurements can help identify issues such as short circuits, open circuits, or degraded insulation within the motor windings.

393. Which of the following maintenance tasks is essential for ensuring the long-term performance and reliability of an electric motor?
a) Lubricating bearings
b) Inspecting and tightening electrical connections
c) Cleaning motor windings and cooling vents
d) All of the above

Answer: d) All of the above
Explanation: Regular maintenance is critical for the long-term performance and reliability of an electric motor. Essential tasks include lubricating bearings, inspecting and tightening electrical connections, and cleaning motor windings and cooling vents to ensure proper cooling and prevent overheating.

394. What is the most common cause of electric motor failure?
a) Bearing failure
b) Winding insulation failure
c) Mechanical overload
d) Voltage imbalance

Answer: a) Bearing failure
Explanation: Bearing failure is the most common cause of electric motor failure. Proper lubrication and maintenance of bearings can help prevent this issue and extend the life of the motor.

395. When repairing an electric motor, which safety precaution should be taken to avoid electrical hazards?
a) Work on the motor while it is energized
b) Disconnect the motor from the power source and verify the absence of voltage
c) Use a metal ladder to reach the motor
d) Use uninsulated tools during the repair process

Answer: b) Disconnect the motor from the power source and verify the absence of voltage
Explanation: When repairing an electric motor, it is crucial to disconnect the motor from the power source and verify the absence of voltage. This safety precaution helps to prevent electrical hazards and ensure the safety of the person performing the repair.

396. Which of the following issues can cause flickering lights in a lighting system?
a) Loose connections
b) Voltage fluctuations
c) Defective lamps or ballasts
d) All of the above

Answer: d) All of the above
Explanation: Flickering lights in a lighting system can be caused by loose connections, voltage fluctuations, or defective lamps or ballasts. Troubleshooting these issues can help identify the root cause and determine the appropriate repairs.

397. What is the first step in troubleshooting a burnt-out lamp?
a) Replace the lamp
b) Check the circuit breaker
c) Inspect the fixture wiring
d) Test the voltage at the lamp socket

Answer: a) Replace the lamp
Explanation: The first step in troubleshooting a burnt-out lamp is to replace it with a known working lamp. If the new lamp does not work, further troubleshooting is needed, such as checking the circuit breaker, inspecting the fixture wiring, or testing the voltage at the lamp socket.

398. When a lighting fixture fails to turn on, which component should be checked first?
a) The lamp
b) The fixture's wiring
c) The wall switch
d) The circuit breaker

Answer: a) The lamp
Explanation: When a lighting fixture fails to turn on, the first component to check is the lamp. If the lamp is not burnt out, then the next steps would be to inspect the wall switch, fixture wiring, and the circuit breaker for any issues.

399. What is the most likely cause of a malfunctioning light fixture that has already had the lamp replaced and the circuit breaker checked?
a) A defective ballast
b) A loose wire connection
c) A faulty wall switch
d) Voltage fluctuations

Answer: b) A loose wire connection
Explanation: If a malfunctioning light fixture still doesn't work after replacing the lamp and checking the circuit breaker, the most likely cause is a loose wire connection within the fixture. Inspecting the fixture's wiring and connections can help identify and fix the issue.

400. What should be done when a lighting system experiences frequent voltage fluctuations?
a) Replace all the lamps
b) Install a voltage regulator
c) Check for loose connections in the wiring
d) Replace the light fixtures

Answer: b) Install a voltage regulator
Explanation: When a lighting system experiences frequent voltage fluctuations, installing a voltage regulator can help stabilize the voltage and prevent issues such as flickering lights and premature lamp failures. Additionally, it is essential to inspect the wiring and connections to ensure there are no loose connections contributing to the fluctuations.

401. A commercial building is experiencing frequent tripping of circuit breakers, and the facility manager has called an electrician to investigate. What should be the electrician's first step in diagnosing the issue?
a) Replace all circuit breakers
b) Identify the circuits experiencing frequent tripping
c) Install larger capacity circuit breakers
d) Check the building's grounding system

Answer: b) Identify the circuits experiencing frequent tripping
Explanation: The electrician should first identify the circuits experiencing frequent tripping to isolate the issue and then investigate potential causes, such as overloaded circuits, short circuits, or faulty breakers.

402. An electrician is called to troubleshoot a lighting system in a residential home, where multiple rooms have flickering lights. What is the most likely cause of this issue?
a) Defective lamps or ballasts in each room
b) Loose connections in individual fixtures
c) Voltage fluctuations affecting the entire home
d) Malfunctioning switches in each room

Answer: c) Voltage fluctuations affecting the entire home
Explanation: If multiple rooms have flickering lights, the most likely cause is voltage fluctuations affecting the entire home. The electrician should investigate potential causes of the fluctuations and implement appropriate solutions.

403. During the installation of a new electrical panel in a commercial building, the electrician discovers that the grounding electrode conductor is undersized. What should the electrician do?
a) Ignore the issue since the panel is already installed
b) Replace the grounding electrode conductor with the correct size
c) Install a second grounding electrode conductor to compensate for the undersized conductor
d) Report the issue to the building owner and leave it up to them to decide

Answer: b) Replace the grounding electrode conductor with the correct size
Explanation: The electrician should replace the undersized grounding electrode conductor with the correct size to ensure proper grounding and safety of the electrical system.

404. A homeowner complains about an electrical outlet that stopped working. The electrician finds that the outlet is properly wired, but the circuit breaker has tripped. After resetting the breaker, the outlet still does not work. What is the most likely cause of this issue?
a) A faulty outlet
b) A short circuit in the wiring
c) A defective circuit breaker
d) The circuit is overloaded

Answer: c) A defective circuit breaker
Explanation: Since the outlet is properly wired and the circuit breaker has been reset, the most likely cause is a defective circuit breaker. The electrician should test the breaker and replace it if necessary.

405. An electrician is troubleshooting a motor that keeps overheating and shutting down. They have already checked the motor for proper voltage and current ratings and ensured that the motor is not overloaded. What is the next step the electrician should take?
a) Replace the motor
b) Inspect and clean the motor's cooling system
c) Check the motor bearings for wear or damage
d) Install a larger capacity motor

Answer: b) Inspect and clean the motor's cooling system
Explanation: The electrician should inspect and clean the motor's cooling system to ensure proper airflow and cooling, as this can cause overheating and shutdowns if not functioning correctly. If the issue persists, further investigation into motor bearings and other potential causes should be conducted.

406. An electrician is designing a new circuit for a residential property, with several outlets and lighting fixtures. The total load for the circuit is 2,800 watts. If the local electrical code requires a continuous load to be no more than 80% of the circuit's capacity, what size breaker should the electrician install for this circuit?
a) 15-amp breaker
b) 20-amp breaker
c) 30-amp breaker
d) 40-amp breaker

Answer: c) 30-amp breaker
Explanation: Circuit capacity = Total Load / 80% = 2,800W / 0.8 = 3,500W. Since the voltage for a residential property is usually 120V, the required amperage would be 3,500W / 120V = 29.17A. Therefore, a 30-amp breaker should be installed.

407. A commercial building has an electrical panel with a 200-amp main breaker. If the building's load calculation indicates a total demand load of 180 amps, can the electrician add an additional 20-amp circuit to the panel?
a) Yes, the panel can handle the additional load
b) No, the panel is already at its maximum capacity
c) Yes, but only if the main breaker is upgraded
d) No, because the additional load would exceed the 80% rule

Answer: a) Yes, the panel can handle the additional load
Explanation: Since the total demand load is 180 amps, the panel can handle an additional 20-amp circuit without exceeding its 200-amp capacity.

408. An electrician is installing a 240V, 4,800W water heater in a residential property. What size wire should be used for the installation?
a) 10 AWG
b) 8 AWG
c) 6 AWG
d) 4 AWG

Answer: c) 6 AWG
Explanation: First, calculate the current: I = P/V = 4,800W / 240V = 20A. Then, consult the National Electrical Code (NEC) or local codes to find the appropriate wire size for 20A, which is typically 6 AWG.

409. A homeowner is considering the installation of a 5kW solar panel system. If the average daily energy consumption of the home is 30 kWh, what percentage of the home's daily energy needs would the solar panel system cover?
a) 50%
b) 66.67%
c) 75%
d) 83.33%

Answer: d) 83.33%
Explanation: The solar panel system would generate 5kW * 24 hours = 120 kWh per day. The percentage of the home's daily energy needs covered would be (120 kWh / 30 kWh) * 100% = 83.33%.

410. A three-phase induction motor has a power factor of 0.85 and an efficiency of 90%. If the motor is rated at 25 HP, what is the total apparent power consumed by the motor in kVA?
a) 25 kVA
b) 29.41 kVA
c) 32.94 kVA
d) 36.47 kVA

Answer: b) 29.41 kVA
Explanation: First, find the real power consumed by the motor: P = HP * 746W/HP / Efficiency = 25 * 746 / 0.9 ≈ 20,684W or 20.68kW. Then, calculate the apparent power: S = P / Power Factor = 20.68kW /

**Case Study. Residential wiring and circuit design.**
**John is an electrician and has been hired to design the wiring and circuit plan for a new residential construction project. The house is a two-story, 4-bedroom home with 3 bathrooms, a kitchen, a dining room, and a living room. John needs to consider the number of circuits required, the appropriate wire sizes, and circuit breaker ratings.**

411. John needs to determine the appropriate number of branch circuits for lighting and receptacle loads in the house. What should he consider while making this decision?
a. The number of receptacles in each room.
b. The total square footage of the house.
c. The number of electrical devices in each room.
d. The total number of rooms in the house.

Answer: b. The total square footage of the house.
Explanation: The total square footage is used to determine the required number of branch circuits for lighting and receptacle loads. This ensures that the electrical system is designed to handle the overall load of the house.

412. The kitchen requires dedicated circuits for appliances such as the refrigerator, dishwasher, and microwave. What type of circuit breakers should John use for these circuits?
a. Standard circuit breakers
b. Ground fault circuit interrupter (GFCI) breakers
c. Arc fault circuit interrupter (AFCI) breakers
d. Combination AFCI/GFCI breakers

Answer: d. Combination AFCI/GFCI breakers
Explanation: Combination AFCI/GFCI breakers provide protection against both arc faults and ground faults, ensuring the highest level of safety for the dedicated circuits in the kitchen.

413. While designing the wiring plan for the bedrooms, John needs to ensure that the wire size is appropriate for the load. If the anticipated load for the bedroom circuits is 15A, which wire size should he use?
a. 12 AWG
b. 14 AWG
c. 10 AWG
d. 8 AWG

Answer: b. 14 AWG
Explanation: For a 15A load, 14 AWG wire size is appropriate. It is capable of safely carrying up to 15A of current, making it suitable for the bedroom circuits.

414. The bathrooms in the house require GFCI protection for receptacles. What is the purpose of GFCI protection?
a. To protect against overcurrents
b. To protect against ground faults
c. To protect against short circuits
d. To protect against voltage fluctuations

Answer: b. To protect against ground faults
Explanation: Ground fault circuit interrupters (GFCIs) protect against ground faults by quickly disconnecting power in case of a ground fault, which can help prevent electrical shocks.

415. John needs to determine the appropriate circuit breaker rating for the water heater in the house, which has a rated power of 4500 watts and operates at 240 volts. What should be the minimum rating for the circuit breaker?
a. 15 A
b. 20 A
c. 25 A
d. 30 A

Answer: d. 30 A
Explanation: To determine the minimum circuit breaker rating, divide the rated power by the operating voltage (4500W / 240V = 18.75A). It's best to select the next highest standard breaker size, which is 30A in this case.

Case Study.
Susan is an experienced electrician tasked with designing the electrical system for a new commercial office building. The building will have multiple floors, each with office spaces, restrooms, a break room, and a conference room. Susan needs to consider factors such as load calculations, branch circuits, and safety requirements.

416. In order to determine the appropriate size of the electrical service for the building, what should Susan calculate first?
a. Total connected load
b. Total demand load
c. Maximum demand load
d. Total utility load

Answer: a. Total connected load
Explanation: Susan should first calculate the total connected load, which is the sum of all the loads connected to the electrical system. This information will help her determine the appropriate size of the electrical service for the building.

417. For the office spaces, Susan needs to ensure that there are enough receptacle outlets to accommodate the needs of the employees. What is the maximum distance between receptacle outlets according to the National Electrical Code (NEC)?
a. 6 feet
b. 12 feet
c. 18 feet
d. 24 feet

Answer: b. 12 feet
Explanation: According to the NEC, receptacle outlets in office spaces should be installed so that no point along the floor line is more than 6 feet from an outlet, which results in a maximum distance of 12 feet between receptacle outlets.

418. In the break rooms, Susan needs to install GFCI protection for the receptacles. What is the main purpose of GFCI protection?
a. To protect against overcurrents
b. To protect against ground faults
c. To protect against short circuits
d. To protect against voltage fluctuations

Answer: b. To protect against ground faults
Explanation: Ground fault circuit interrupters (GFCIs) protect against ground faults by quickly disconnecting power in case of a ground fault, which can help prevent electrical shocks.

419. Susan needs to determine the appropriate type of conduit to use for the electrical wiring in the building. Which type of conduit is best suited for commercial installations with exposed wiring?
a. PVC conduit
b. Flexible metallic conduit (FMC)
c. Rigid metal conduit (RMC)
d. Electrical metallic tubing (EMT)

Answer: d. Electrical metallic tubing (EMT)
Explanation: Electrical metallic tubing (EMT) is a common choice for commercial installations with exposed wiring because it provides a balance of durability and ease of installation.

420. The lighting system in the building needs to be designed for energy efficiency. Which type of lighting control is most suitable for reducing energy consumption in the office spaces?
a. Timers
b. Motion sensors
c. Dimmers
d. Photocells

Answer: b. Motion sensors
Explanation: Motion sensors can help reduce energy consumption by automatically turning off the lights when no movement is detected in a space, ensuring that lights are not left on unnecessarily.

Case Study.
John is an experienced electrician who has been hired to perform an electrical service and panel upgrade for a 30-year-old residential property. The homeowner wants to install a central air conditioning system, an electric vehicle charging station, and modern appliances. John needs to assess the current electrical service, determine the necessary upgrades, and ensure compliance with the National Electrical Code (NEC).

421. What is the first step John should take when evaluating the current electrical service?
a. Calculate the current total connected load
b. Inspect the existing electrical panel
c. Check the service entrance conductors
d. Review the grounding and bonding system

Answer: b. Inspect the existing electrical panel
Explanation: John should first inspect the existing electrical panel to assess its condition, identify any potential issues, and determine if it can accommodate the additional loads from the new appliances and systems.

422. Which formula should John use to calculate the minimum required service size for the upgraded electrical system?
a. Service size = Total connected load x 125%
b. Service size = Total demand load x 125%
c. Service size = Total connected load + 25% of the largest load
d. Service size = Total demand load + 25% of the largest load

Answer: b. Service size = Total demand load x 125%
Explanation: According to the NEC, the minimum required service size should be calculated by multiplying the total demand load by 125%. This accounts for both continuous and non-continuous loads.

423. When upgrading the electrical panel, John needs to ensure that it is located in a suitable area. Which of the following locations is NOT appropriate for an electrical panel?
a. In a utility room
b. In a hallway
c. In a bathroom
d. In a garage

Answer: c. In a bathroom
Explanation: According to the NEC, electrical panels should not be installed in bathrooms due to the increased risk of electrical shock in wet environments.

424. John discovers that the existing grounding electrode system is inadequate. Which of the following grounding electrodes is most commonly used for residential service upgrades?
a. Grounding plate electrode
b. Grounding ring electrode
c. Concrete-encased electrode
d. Ground rod electrode

Answer: d. Ground rod electrode
Explanation: Ground rod electrodes are the most commonly used grounding electrodes for residential service upgrades due to their simplicity, cost-effectiveness, and ease of installation.

425. After upgrading the electrical service and panel, what is the most appropriate method for John to ensure that the new installation is safe and compliant with the NEC?
a. Perform a visual inspection of the installation
b. Test the system using a multimeter
c. Obtain an inspection from the local electrical inspector
d. Verify that the installation meets the homeowner's requirements

Answer: c. Obtain an inspection from the local electrical inspector
Explanation: The most appropriate method to ensure that the new installation is safe and compliant with the NEC is to obtain an inspection from the local electrical inspector. This will help identify any potential issues and confirm that the installation meets all code requirements.

Case Study.
Samantha is a licensed electrician working on a new commercial building project. She has been tasked with designing and installing a proper grounding and bonding system for the building's electrical system. Samantha needs to ensure that the system is safe, effective, and compliant with the National Electrical Code (NEC).

426. What is the primary purpose of grounding and bonding in electrical systems?
a. To prevent electrical fires
b. To ensure proper functioning of electrical devices
c. To provide a path for fault current to safely return to the source
d. To reduce the risk of electrocution

Answer: c. To provide a path for fault current to safely return to the source
Explanation: The primary purpose of grounding and bonding is to provide a low-impedance path for fault current to flow back to the source, which helps clear the fault and minimize the risk of electrical fires and electrocution.

427. According to the NEC, what is the minimum size of a grounding electrode conductor for a 200-ampere service using copper conductors?
a. 6 AWG
b. 4 AWG
c. 2 AWG
d. 1/0 AWG

Answer: b. 4 AWG
Explanation: The NEC requires a minimum size of 4 AWG copper conductor for a 200-ampere service grounding electrode conductor.

428. Which of the following is NOT considered an acceptable grounding electrode according to the NEC?
a. Ground rod electrode
b. Concrete-encased electrode
c. Metallic water pipe electrode
d. Aluminum conduit electrode

Answer: d. Aluminum conduit electrode
Explanation: The NEC does not consider aluminum conduit as an acceptable grounding electrode due to its potential for corrosion and lack of effective grounding properties.

429. When bonding metal parts of an electrical system, what is the primary purpose of using a bonding jumper?
a. To provide a low-impedance path for fault current
b. To reduce the resistance of the grounding electrode conductor
c. To protect against corrosion at the bonding connections
d. To ensure proper functioning of electrical devices

Answer: a. To provide a low-impedance path for fault current
Explanation: Bonding jumpers are used to provide a low-impedance path for fault current between metal parts of an electrical system, ensuring that the fault current returns to the source and minimizing the risk of electrical fires and electrocution.

430. When installing a grounding electrode system for a commercial building, what is the minimum required depth for burying a ground rod electrode?
a. 6 feet
b. 4 feet
c. 8 feet
d. 10 feet

Answer: c. 8 feet
Explanation: The NEC requires a minimum burial depth of 8 feet for ground rod electrodes to ensure proper grounding and reduce the risk of corrosion or accidental damage.

Case Study.
John is an experienced electrician who has been hired to install and configure circuit protection devices for a new commercial building. The building requires a variety of fuses and circuit breakers for different applications. John must ensure that the devices are correctly sized and rated according to the National Electrical Code (NEC) and the specific needs of the building.

431. What is the primary purpose of circuit protection devices in electrical systems?
a. To prevent electrical fires
b. To reduce the risk of electrocution
c. To protect electrical equipment from damage due to excessive current
d. To control the flow of electricity

Answer: c. To protect electrical equipment from damage due to excessive current
Explanation: Circuit protection devices, such as fuses and circuit breakers, are designed to protect electrical equipment from damage by interrupting the flow of excessive current caused by short circuits, overloads, or other faults.

432. According to the NEC, what is the maximum allowable continuous current for a 20-ampere circuit breaker?
a. 16 amperes
b. 18 amperes
c. 20 amperes
d. 24 amperes

Answer: a. 16 amperes
Explanation: The NEC requires that circuit breakers be sized to carry no more than 80% of their rated current for continuous loads. For a 20-ampere circuit breaker, this means a maximum allowable continuous current of 16 amperes (20 x 0.8 = 16).

433. When selecting a fuse for a specific application, which of the following factors should be considered?
a. Amperage rating
b. Voltage rating
c. Interrupting rating
d. All of the above

Answer: d. All of the above
Explanation: When selecting a fuse, it is important to consider its amperage rating, voltage rating, and interrupting rating to ensure that the fuse is appropriate for the specific application and can effectively protect the electrical equipment.

434. What is the primary difference between a time-delay fuse and a fast-acting fuse?
a. Time-delay fuses have a higher amperage rating
b. Fast-acting fuses are more sensitive to short circuits
c. Time-delay fuses can tolerate short-duration overcurrent events without opening
d. Fast-acting fuses have a higher voltage rating

Answer: c. Time-delay fuses can tolerate short-duration overcurrent events without opening
Explanation: Time-delay fuses are designed to tolerate short-duration overcurrent events, such as motor inrush currents or temporary load surges, without opening. Fast-acting fuses, on the other hand, open more quickly in response to overcurrent events and provide more sensitive protection for delicate electronic equipment.

435. What is the primary purpose of a ground-fault circuit interrupter (GFCI) in an electrical system?
a. To protect against electrical fires
b. To reduce the risk of electrocution
c. To protect electrical equipment from damage due to excessive current
d. To control the flow of electricity

Answer: b. To reduce the risk of electrocution
Explanation: GFCIs are designed to quickly detect and interrupt the flow of electricity when a ground fault occurs, significantly reducing the risk of electrocution. They are typically used in areas where there is an increased risk of electrical shock, such as bathrooms, kitchens, and outdoor receptacles.

Case Study.
Jane, a licensed electrician, is working on a project that involves wiring a new office building. The project requires her to use various wiring methods and techniques for different applications, such as lighting, power outlets, and HVAC systems. She must ensure that the wiring methods used comply with the National Electrical Code (NEC) and meet the specific needs of each application.

436. Which of the following wiring methods is commonly used for exposed wiring in commercial and industrial applications due to its mechanical protection and ease of installation?
a. Nonmetallic sheathed cable (NM-B)
b. Electrical metallic tubing (EMT)
c. Armored cable (AC)
d. Flexible metal conduit (FMC)

Answer: b. Electrical metallic tubing (EMT)
Explanation: EMT is a commonly used wiring method for exposed wiring in commercial and industrial applications because it provides mechanical protection for the conductors and is relatively easy to install. It can be bent to accommodate various routing requirements and is often used for branch circuit wiring.

437. What type of cable is commonly used for wiring residential circuits and is characterized by its plastic outer sheathing and multiple insulated conductors?
a. Nonmetallic sheathed cable (NM-B)
b. Electrical metallic tubing (EMT)
c. Armored cable (AC)
d. Flexible metal conduit (FMC)

Answer: a. Nonmetallic sheathed cable (NM-B)
Explanation: Nonmetallic sheathed cable (NM-B) is commonly used for wiring residential circuits. It consists of multiple insulated conductors enclosed within a plastic outer sheathing. It is relatively easy to install and is typically used for general-purpose wiring, such as lighting and power outlets.

438. In which of the following applications would you typically use a liquidtight flexible metal conduit (LFMC)?
a. Outdoor installations exposed to moisture
b. Indoor installations with tight bends
c. Concealed wiring in drywall
d. Exposed wiring in commercial buildings

Answer: a. Outdoor installations exposed to moisture
Explanation: LFMC is a flexible metallic conduit with a liquidtight outer jacket, making it suitable for outdoor installations exposed to moisture or corrosive environments. It provides mechanical protection for the conductors and can be used in various applications, including wiring for air conditioning units, pumps, and other outdoor equipment.

439. When connecting two lengths of conduit, which type of fitting should be used to ensure a secure and continuous raceway for the conductors?
a. Conduit clamp
b. Conduit strap
c. Conduit coupling
d. Conduit bushing

Answer: c. Conduit coupling
Explanation: Conduit couplings are used to connect two lengths of conduit and provide a secure and continuous raceway for the conductors. They are available in various materials and types to match the specific conduit being used, such as threaded or compression couplings for rigid metal conduit (RMC) or set-screw couplings for electrical metallic tubing (EMT).

440. In order to maintain the integrity of a fire-rated wall, which of the following should be used to seal around a conduit penetration?
a. Duct seal compound
b. Firestop sealant
c. Conduit bushing
d. Conduit locknut

Answer: b. Firestop sealant
Explanation: Firestop sealant is used to seal around conduit penetrations in fire-rated walls, floors, and ceilings. It is designed to expand when exposed to high temperatures, effectively sealing the penetration and preventing the spread of fire and smoke through the opening. Using firestop sealant helps maintain the integrity of the fire-rated assembly and is required by the NEC and building codes.

Case Study.
John, an experienced electrician, is working on a residential project that involves installing various electrical devices, including receptacles, switches, and light fixtures. He must ensure that the installation complies with the National Electrical Code (NEC) and follows best practices for safety and performance.

441. When installing a receptacle in a kitchen, what is the minimum distance that must be maintained between the receptacle and the edge of the kitchen sink?
a. 6 inches
b. 12 inches
c. 18 inches
d. 24 inches

Answer: b. 12 inches. Explanation: According to the NEC, receptacles installed in a kitchen must be at least 12 inches from the edge of the kitchen sink. This helps to minimize the risk of electrical shock by keeping the receptacle away from water sources.

442. When installing a switch in a residential setting, what is the recommended mounting height for the switch from the finished floor?
a. 36 inches
b. 42 inches
c. 48 inches
d. 54 inches

Answer: c. 48 inches
Explanation: The recommended mounting height for switches in residential settings is 48 inches from the finished floor to the center of the switch box. This height is considered comfortable and accessible for most users.

443. When wiring a three-way switch, which of the following conductors is used to connect the two switches together?
a. Ground wire
b. Neutral wire
c. Hot wire
d. Traveler wire

Answer: d. Traveler wire
Explanation: Traveler wires are used to connect two three-way switches together in a circuit. They allow the switches to control a light or other load from two different locations. Two traveler wires are typically used in a three-way switch setup, one for each possible switch position.

444. When installing a light fixture, what should be done with the ground wire from the fixture if the electrical box is metal and already grounded?
a. Leave the ground wire unconnected
b. Connect the ground wire to the metal box
c. Connect the ground wire to the neutral wire
d. Connect the ground wire to the hot wire

Answer: b. Connect the ground wire to the metal box
Explanation: When installing a light fixture in a metal electrical box that is already grounded, the ground wire from the fixture should be connected to the metal box using a grounding screw or a grounding clip. This ensures that the fixture is properly grounded and helps to minimize the risk of electrical shock.

445. When installing a GFCI receptacle in a bathroom, which of the following wiring configurations is correct?
a. Connect the load terminals to the incoming power source and the line terminals to the downstream receptacles
b. Connect the line terminals to the incoming power source and the load terminals to the downstream receptacles
c. Connect both the line and load terminals to the incoming power source
d. Connect both the line and load terminals to the downstream receptacles

Answer: b. Connect the line terminals to the incoming power source and the load terminals to the downstream receptacles

Explanation: When installing a GFCI receptacle, the line terminals should be connected to the incoming power source, and the load terminals should be connected to the downstream receptacles that require GFCI protection. This configuration ensures that the GFCI receptacle provides the necessary protection for the downstream receptacles.

Case Study.

Emma, an electrician, is called to troubleshoot various electrical issues at a residential property. She has to identify the root causes of the problems and fix them, ensuring that the electrical system is safe and functional. She uses her knowledge and tools, such as multimeters and clamp meters, to diagnose and repair the issues.

446. Emma is investigating a circuit that is experiencing intermittent power loss. What is the most likely cause of this issue?
a. Open circuit
b. Short circuit
c. Ground fault
d. Overloaded circuit

Answer: a. Open circuit

Explanation: Intermittent power loss is often caused by an open circuit, which occurs when there is a break or disconnection in the circuit. This can be due to loose connections, damaged wires, or faulty devices.

447. During her investigation, Emma finds a circuit breaker that keeps tripping. What is the most likely cause of this issue?
a. Open circuit
b. Short circuit
c. Ground fault
d. Overloaded circuit

Answer: d. Overloaded circuit

Explanation: A circuit breaker that keeps tripping is most likely experiencing an overloaded circuit. An overloaded circuit occurs when the current flowing through the circuit exceeds the rated capacity of the circuit breaker or fuse, causing it to trip or blow to protect the circuit from overheating and potential fire hazards.

448. Emma suspects that a ground fault is causing a GFCI outlet to trip frequently. What tool should she use to confirm her suspicion?
a. Multimeter
b. Clamp meter
c. Continuity tester
d. Voltage tester

Answer: a. Multimeter
Explanation: A multimeter can be used to measure resistance between the hot wire and ground, which will help Emma determine if there is a ground fault causing the GFCI outlet to trip. If the resistance reading is low, it indicates that there is a ground fault present.

449. While troubleshooting a non-functioning light fixture, Emma finds that the voltage at the fixture is significantly lower than the expected 120 volts. What could be the cause of this issue?
a. Open circuit
b. Short circuit
c. Ground fault
d. Voltage drop

Answer: d. Voltage drop
Explanation: A significant voltage drop at the light fixture could be caused by factors such as long wire runs, undersized wires, or poor connections. Voltage drop can result in dim or non-functioning lights, as well as reduced performance and efficiency of electrical devices.

450. Emma encounters a situation where a motor keeps overheating and shutting down. What is the most likely cause of this issue?
a. Open circuit
b. Short circuit
c. Ground fault
d. Overloaded motor

Answer: d. Overloaded motor
Explanation: An overheating motor that keeps shutting down is likely experiencing an overload. Overloading can occur when the motor is forced to operate beyond its rated capacity, resulting in excessive heat generation and, eventually, motor shutdown to protect it from damage. Causes of motor overloading can include mechanical issues, excessive load, or improper wiring.

Case Study.
James, an electrician, is hired to perform preventive maintenance on a commercial building's electrical system. His tasks include inspecting connections, testing devices, and ensuring that the system is functioning safely and efficiently. He must identify any potential issues and make necessary repairs to prevent future problems.

451. During his inspection, James notices some discolored and corroded connections in an electrical panel. What is the most likely cause of this issue?
a. Overloaded circuit
b. Voltage drop
c. Loose connections
d. Moisture intrusion

Answer: d. Moisture intrusion
Explanation: Discolored and corroded connections are often caused by moisture intrusion, which can lead to poor connections, overheating, and potential electrical hazards. Moisture can enter the panel through leaks, condensation, or high humidity levels.

452. James wants to test the functionality of a ground fault circuit interrupter (GFCI) outlet. What should he use to perform this test?
a. Multimeter
b. Clamp meter
c. Continuity tester
d. GFCI tester

Answer: d. GFCI tester
Explanation: A GFCI tester is specifically designed to test the functionality of GFCI outlets by simulating a ground fault, causing the GFCI to trip. This ensures that the GFCI is functioning properly and providing the necessary protection against electrical shock hazards.

453. During his inspection, James finds a circuit breaker that is hot to the touch. What should be his primary concern in this situation?
a. Open circuit
b. Short circuit
c. Overloaded circuit
d. Ground fault

Answer: c. Overloaded circuit
Explanation: A hot circuit breaker is often a sign of an overloaded circuit. An overloaded circuit occurs when the current flowing through the circuit exceeds the rated capacity of the circuit breaker, causing it to overheat. This can lead to potential fire hazards and equipment damage.

454. James notices flickering lights in a particular area of the building. What is the most likely cause of this issue?
a. Open circuit
b. Loose connections
c. Ground fault
d. Voltage drop

Answer: b. Loose connections
Explanation: Flickering lights are often caused by loose connections in the circuit, which can lead to intermittent power disruptions. Loose connections can result from poor installation, vibration, or thermal expansion and contraction.

455. As part of his preventive maintenance tasks, James must ensure that all electrical connections are properly tightened. What tool should he use to achieve this?
a. Multimeter
b. Torque wrench
c. Continuity tester
d. Wire stripper

Answer: b. Torque wrench
Explanation: A torque wrench is used to apply the correct amount of torque to electrical connections, ensuring that they are properly tightened and secure. Properly torqued connections reduce the risk of overheating, arcing, and other electrical hazards.

Case Study.
Linda, an experienced electrician, is called to a commercial building to investigate issues with the building's transformers. Her task is to identify the root cause of the problems, repair or replace the faulty transformers, and ensure that the electrical system is functioning properly.

456. Linda discovers that one of the transformers is overheating. What is the most likely cause of this issue?
a. Overload
b. Short circuit
c. Loose connections
d. Incorrect transformer rating

Answer: a. Overload
Explanation: Overheating in transformers is often caused by an overload, which occurs when the transformer is subjected to a higher load than its rated capacity. Overheating can lead to insulation failure, reduced transformer life, and other safety hazards.

457. While troubleshooting, Linda finds that the secondary voltage of a transformer is lower than expected. What could be the cause of this issue?
a. Excessive load
b. Open secondary winding
c. Shorted primary winding
d. Incorrect tap settings

Answer: d. Incorrect tap settings
Explanation: Incorrect tap settings can cause the secondary voltage to be lower than expected. Taps are used to adjust the output voltage of a transformer to accommodate varying load requirements or compensate for voltage drops in the distribution system.

458. Linda suspects that a transformer has a shorted winding. Which testing method can help her confirm her suspicion?
a. Insulation resistance test
b. Winding resistance test
c. Transformer turns ratio test
d. Polarity test

Answer: b. Winding resistance test
Explanation: A winding resistance test can help determine if there is a shorted winding in a transformer. The test involves measuring the resistance of the transformer windings and comparing the measurements to the manufacturer's specifications or similar transformers in service.

459. While investigating transformer issues, Linda discovers that one of the transformers has a failed insulation. What should she do in this situation?
a. Repair the insulation
b. Replace the transformer
c. Adjust the tap settings
d. Reconnect the transformer

Answer: b. Replace the transformer. Explanation: A transformer with failed insulation should be replaced, as repairing the insulation may not be reliable or cost-effective. Failed insulation can lead to short circuits, electrical hazards, and equipment damage.

460. Linda needs to select a replacement transformer for a failed unit. Which of the following factors should she consider when making her selection?
a. Transformer type and application
b. Voltage and current ratings
c. Impedance and efficiency
d. All of the above

Answer: d. All of the above
Explanation: When selecting a replacement transformer, Linda should consider the transformer type and its intended application, voltage and current ratings, and impedance and efficiency. These factors ensure that the replacement transformer will function properly and meet the requirements of the electrical system.

Case Study.
Jack, an experienced electrician, has been called to a manufacturing facility to investigate and repair issues with several electric motors. His task is to identify the causes of the motor problems, repair or replace the faulty components, and ensure that the motors are functioning properly.

461. Jack discovers that a motor is drawing excessive current. What could be the cause of this issue?
a. Voltage imbalance
b. Worn bearings
c. Incorrect motor connections
d. Ground fault

Answer: a. Voltage imbalance
Explanation: Voltage imbalance can cause a motor to draw excessive current, which can lead to overheating and reduced motor life. Imbalances can result from an unbalanced power source, unevenly distributed loads, or poor connections in the distribution system.

462. While troubleshooting, Jack finds that a motor is not starting but is humming. What is the most likely cause of this issue?
a. Open motor winding
b. Locked rotor
c. Blown fuse
d. Faulty capacitor

Answer: d. Faulty capacitor
Explanation: A humming motor that fails to start is often indicative of a faulty capacitor. Capacitors are used in single-phase motors to create a phase shift, which helps the motor start. A faulty capacitor can prevent the motor from starting or cause it to run inefficiently.

463. Jack notices that a motor is overheating during operation. Which of the following could be a possible cause?
a. Undervoltage
b. Overvoltage
c. Misalignment
d. All of the above

Answer: d. All of the above
Explanation: Overheating in motors can be caused by various factors, including undervoltage, overvoltage, and misalignment. These issues can lead to excessive current draw, increased bearing temperatures, and reduced motor life.

464. When investigating a motor that trips the circuit breaker upon starting, Jack suspects a short circuit. Which testing method can help him confirm his suspicion?
a. Insulation resistance test
b. Winding resistance test
c. Megger test
d. Motor rotation test

Answer: a. Insulation resistance test
Explanation: An insulation resistance test can help determine if there is a short circuit in a motor. The test involves applying a high voltage to the motor windings and measuring the resistance of the insulation. A low resistance reading indicates a short circuit.

465. Jack needs to replace a motor's bearings. What should he consider when selecting the appropriate replacement bearings?
a. Bearing type and size
b. Operating temperature range
c. Load capacity
d. All of the above

Answer: d. All of the above
Explanation: When selecting replacement bearings, Jack should consider the bearing type and size, operating temperature range, and load capacity. These factors ensure that the replacement bearings will function properly and meet the requirements of the motor.

Case Study.
Anna, a seasoned electrician, is working on a project to maintain and troubleshoot lighting systems in a large office building. She needs to diagnose various lighting issues and ensure that the lighting system is functioning optimally.

466. While inspecting the building's lighting system, Anna discovers a flickering light. What is the most likely cause of this issue?
a. Burnt-out lamp
b. Loose wiring connections
c. Faulty ballast
d. Damaged fixture

Answer: b. Loose wiring connections
Explanation: Flickering lights are often caused by loose wiring connections. This issue can lead to intermittent power supply to the lamp, causing it to flicker. Anna should check and tighten the wiring connections to resolve the issue.

467. Anna encounters a buzzing sound coming from a fluorescent light fixture. Which component is most likely responsible for the noise?
a. Lamp
b. Ballast
c. Fixture
d. Capacitor

Answer: b. Ballast
Explanation: Buzzing sounds in fluorescent light fixtures are often due to a faulty or aging ballast. As the ballast begins to wear out, it can produce a buzzing noise. Replacing the ballast should resolve the issue.

468. When a light switch is turned on, the corresponding light does not illuminate. What should Anna check first to diagnose the issue?
a. The light switch
b. The wiring connections
c. The fixture
d. The lamp

Answer: d. The lamp
Explanation: The first step in diagnosing a non-functioning light is to check the lamp. If the lamp is burnt out or damaged, replacing it should resolve the issue. If the lamp is in good condition, Anna should proceed to check other components such as the switch, wiring connections, and fixture.

469. Anna is troubleshooting a malfunctioning LED light fixture. What is a common issue that may cause an LED fixture to fail prematurely?
a. Excessive heat
b. Incompatible dimmer switch
c. Incorrect voltage
d. All of the above

Answer: d. All of the above
Explanation: LED light fixtures can fail prematurely due to several factors, including excessive heat, an incompatible dimmer switch, or incorrect voltage. Anna should investigate these potential issues to identify and resolve the problem.

470. While maintaining the lighting system, Anna notices that several lamps have a significantly reduced light output. What is the likely cause of this issue?
a. Dirty lamps or fixtures
b. Voltage fluctuation
c. Aging lamps
d. Both a and c

Answer: d. Both a and c
Explanation: Reduced light output can be caused by dirty lamps or fixtures, as well as aging lamps. Anna should clean the lamps and fixtures to remove any accumulated dust or dirt. If the light output remains low after cleaning, it may be necessary to replace the aging lamps with new ones.

Case Study.
John is an electrical engineer tasked with designing the electrical system for a new commercial building. He must perform load calculations and other electrical calculations to ensure the building meets code requirements and operates efficiently.

471. John needs to calculate the general lighting load for a 2,500 square foot office space. If the lighting load is calculated at 3.5 VA/sq ft, what is the total lighting load?
a. 7,500 VA
b. 8,750 VA
c. 10,000 VA
d. 12,500 VA

Answer: b. 8,750 VA
Explanation: To calculate the lighting load, multiply the office space square footage by the VA per square foot: 2,500 sq ft x 3.5 VA/sq ft = 8,750 VA.

472. The building has a motor with a power rating of 10 HP. If the motor's efficiency is 90%, what is its input power in kilowatts (kW)?
a. 7.46 kW
b. 8.29 kW
c. 9.0 kW
d. 10.0 kW

Answer: a. 7.46 kW
Explanation: First, convert the motor's power rating from HP to watts: 10 HP x 746 W/HP = 7,460 W. Then, divide the output power by the motor's efficiency: 7,460 W / 0.9 = 8,289 W. Finally, convert to kilowatts: 8,289 W / 1,000 W/kW = 8.29 kW.

473. John needs to determine the appropriate wire size for a 60-ampere circuit with a maximum voltage drop of 3%. If the one-way distance is 100 feet and the wire's resistance is 0.1 ohms per 1,000 feet, what wire size should he choose?
a. 4 AWG
b. 6 AWG
c. 8 AWG
d. 10 AWG

Answer: b. 6 AWG
Explanation: To calculate the maximum allowed voltage drop, multiply the circuit's amperage by the wire's resistance and distance: 60 A x 0.1 Ω/1,000 ft x 100 ft = 0.6 V. Then, divide the voltage drop by the percentage allowed (3%): 0.6 V / 0.03 = 20 V. Using a wire size chart, 6 AWG is the appropriate wire size for a 20 V drop in a 60-ampere circuit.

474. The building requires a feeder with a 120/240 V single-phase service. If the calculated load is 48 kVA, what is the minimum required service amperage?
a. 160 A
b. 200 A
c. 240 A
d. 320 A

Answer: b. 200 A
Explanation: To calculate the minimum required service amperage, divide the calculated load by the service voltage: 48,000 VA / 240 V = 200 A.

475. John needs to calculate the required conductor size for a 100 A, 120 V circuit with a one-way distance of 250 feet and a maximum voltage drop of 3%. If the wire's resistance is 0.1 ohms per 1,000 feet, what conductor size should he choose?
a. 1 AWG
b. 2 AWG
c. 3 AWG
d. 4 AWG

Answer: a. 1 AWG

Case Study.
Jane, an electrician, is working on a commercial building renovation project. She encounters various electrical installations and troubleshooting tasks, and must adhere to safety practices throughout the project.

476. Jane is working on a live electrical panel. What is the minimum level of personal protective equipment (PPE) she should wear?
a. Safety glasses and insulated gloves
b. Safety glasses, insulated gloves, and flame-resistant clothing
c. Safety glasses, insulated gloves, flame-resistant clothing, and arc-rated face shield
d. No PPE is required if the panel is properly labeled

Answer: c. Safety glasses, insulated gloves, flame-resistant clothing, and arc-rated face shield
Explanation: Working on live electrical equipment requires comprehensive PPE, including safety glasses, insulated gloves, flame-resistant clothing, and an arc-rated face shield to minimize the risk of injury.

477. What should Jane do before starting any electrical work on a circuit?
a. Test the circuit with a multimeter
b. Visually inspect the wiring
c. Disconnect the power and lockout/tagout the circuit
d. Consult with other team members

Answer: c. Disconnect the power and lockout/tagout the circuit
Explanation: To ensure safety, Jane must disconnect the power and lockout/tagout the circuit before starting any electrical work to prevent accidental re-energizing.

478. Jane encounters an open junction box with exposed wires. What is the appropriate action to take?
a. Wrap the wires with electrical tape
b. Install a cover on the junction box
c. Report the issue to the supervisor
d. Both b and c

Answer: d. Both b and c
Explanation: The appropriate action is to both install a cover on the junction box to prevent accidental contact with the exposed wires and report the issue to the supervisor to ensure proper follow-up and documentation.

479. While troubleshooting an electrical issue, Jane uses a clamp meter to measure current. What is the safest way to use the clamp meter?
a. Clamp it around a single conductor
b. Clamp it around multiple conductors
c. Insert the clamp meter's probe into the circuit
d. None of the above

Answer: a. Clamp it around a single conductor
Explanation: Clamp meters are designed to measure current safely by clamping around a single conductor, ensuring that there is no direct contact with live electrical parts.

480. Jane needs to work on a high-voltage system. What additional safety measures should she consider?
a. Use insulated tools
b. Use a voltage detector
c. Maintain a safe distance from live components
d. All of the above


Answer: d. All of the above
Explanation: When working on high-voltage systems, Jane should use insulated tools, a voltage detector to ensure circuits are de-energized, and maintain a safe distance from live components to minimize the risk of injury.

Case Study.
Alex, a licensed electrician, is responsible for ensuring that a new residential construction project complies with the National Electrical Code (NEC) and other relevant codes and standards.

481. Alex discovers that the project specifications require the installation of a 20-amp circuit for a bathroom. What is the minimum wire size Alex should use for this circuit?
a. 14 AWG
b. 12 AWG
c. 10 AWG
d. 8 AWG


Answer: b. 12 AWG
Explanation: According to the NEC, a 20-amp circuit requires a minimum wire size of 12 AWG.

482. What is the required minimum spacing between receptacle outlets in a residential kitchen along the countertop?
a. 4 feet
b. 6 feet
c. 8 feet
d. 12 feet


Answer: b. 6 feet
Explanation: The NEC requires receptacle outlets to be spaced no more than 6 feet apart along kitchen countertops to ensure that no point on the countertop is more than 2 feet from an outlet.

483. What is the maximum height for switches and circuit breakers in a residential setting, according to the NEC?
a. 48 inches
b. 60 inches
c. 66 inches
d. 72 inches

Answer: c. 66 inches
Explanation: The NEC specifies that switches and circuit breakers should be installed at a maximum height of 66 inches above the floor in residential settings.

484. Alex needs to install a subpanel in the new residential construction project. What is the minimum working space required in front of the subpanel, according to the NEC?
a. 24 inches
b. 30 inches
c. 36 inches
d. 42 inches

Answer: c. 36 inches
Explanation: The NEC requires a minimum working space of 36 inches in front of electrical panels, including subpanels, to ensure adequate space for safe access and maintenance.

485. According to the NEC, what type of receptacle is required for a residential laundry room?
a. 15-amp duplex receptacle
b. 20-amp duplex receptacle
c. 15-amp ground-fault circuit interrupter (GFCI) receptacle
d. 20-amp ground-fault circuit interrupter (GFCI) receptacle

Answer: d. 20-amp ground-fault circuit interrupter (GFCI) receptacle
Explanation: The NEC requires a 20-amp GFCI receptacle to be installed in residential laundry rooms to provide additional protection against electrical shock.

Case Study
Jane is an electrical contractor tasked with designing and installing an emergency and backup power system for a medium-sized commercial building. The building requires a standby generator and an uninterruptible power supply (UPS) system to maintain power for essential equipment during power outages.

486. What type of transfer switch should Jane use to safely connect the standby generator to the building's electrical system?
a. Manual transfer switch
b. Automatic transfer switch
c. Double-throw transfer switch
d. Single-throw transfer switch

Answer: b. Automatic transfer switch
Explanation: An automatic transfer switch is used to automatically switch the building's electrical load between the utility power and the standby generator during power outages, ensuring a seamless transition and continuous power supply.

487. When designing the backup power system, which type of UPS system should Jane consider for protecting sensitive electronic equipment, such as computers and servers?
a. Standby UPS
b. Line-interactive UPS
c. Double-conversion online UPS
d. Ferroresonant UPS

Answer: c. Double-conversion online UPS
Explanation: A double-conversion online UPS system provides the highest level of power protection by continuously converting incoming AC power to DC power and then back to clean, regulated AC power, isolating sensitive equipment from power fluctuations and disturbances.

488. Which of the following factors should Jane consider when selecting the appropriate size for the standby generator?
a. Building square footage
b. Total connected load
c. Required voltage and frequency
d. Both b and c

Answer: d. Both b and c. Explanation: To size a standby generator correctly, Jane must consider the total connected load that the generator will need to support during a power outage and ensure that the generator can supply the required voltage and frequency.

489. What type of fuel is commonly used for commercial standby generators due to its long storage life and availability?
a. Gasoline
b. Diesel
c. Natural gas
d. Propane

Answer: b. Diesel
Explanation: Diesel fuel is commonly used for commercial standby generators due to its long storage life and widespread availability, making it a reliable choice for emergency power applications.

490. What is a critical factor Jane should consider when designing the UPS system to ensure adequate backup power during an outage?
a. Generator size
b. Battery type
c. UPS system efficiency
d. Battery runtime

Answer: d. Battery runtime
Explanation: When designing a UPS system, ensuring adequate battery runtime is crucial to maintaining power for essential equipment during an outage until the standby generator comes online or the utility power is restored.

Case Study
John is an electrical contractor who has been hired to design and install a solar panel system for a residential property. The homeowners want to offset their electricity usage with solar energy and reduce their reliance on the grid. John must evaluate the property's energy consumption, available roof space, and other factors to design the most effective solar panel system.

491. Which of the following factors is most important when determining the optimal angle and orientation for the solar panels?
a. Roof pitch
b. Prevailing winds
c. Latitude
d. Local weather conditions

Answer: c. Latitude
Explanation: The optimal angle and orientation of solar panels depend on the property's latitude. In general, solar panels should be tilted at an angle equal to the latitude, facing south in the northern hemisphere, and north in the southern hemisphere, to maximize energy production.

492. What type of solar panel mounting system would John recommend for a flat roof with limited space and load-bearing capacity?
a. Ground-mounted system
b. Roof-mounted system with a fixed tilt
c. Roof-mounted system with adjustable tilt
d. Ballasted mounting system

Answer: d. Ballasted mounting system
Explanation: A ballasted mounting system is a non-penetrating, roof-mounted system that uses weights (ballast) to hold the solar panels in place. This type of mounting system is suitable for flat roofs with limited space and load-bearing capacity as it does not require roof penetrations and minimizes additional stress on the roof structure.

493. To calculate the required number of solar panels, which of the following factors should John consider?
a. Total energy consumption of the property
b. Solar panel efficiency
c. Solar irradiance at the location
d. All of the above

Answer: d. All of the above
Explanation: To determine the required number of solar panels, John must consider the total energy consumption of the property, the efficiency of the solar panels, and the solar irradiance at the location. These factors will help him determine the appropriate system size to offset the property's electricity usage.

494. What is the most commonly used type of solar panel for residential installations due to its efficiency and cost-effectiveness?
a. Monocrystalline silicon
b. Polycrystalline silicon
c. Thin-film
d. Concentrated photovoltaic (CPV)

Answer: b. Polycrystalline silicon
Explanation: Polycrystalline silicon solar panels are the most commonly used type of solar panel for residential installations because they offer a good balance between efficiency and cost-effectiveness.

495. Which of the following components is essential for converting the DC power generated by solar panels into AC power that can be used by the home's electrical system?
a. Charge controller
b. Inverter
c. Solar battery
d. Disconnect switch

Answer: b. Inverter
Explanation: An inverter is a critical component of a solar panel system, as it converts the DC power generated by the solar panels into AC power that can be used by the home's electrical system or fed back into the grid.

Case Study
Emily is an electrical engineer who has been hired by a large commercial building owner to improve energy efficiency and reduce energy consumption. She is tasked with assessing the building's electrical systems, including lighting, HVAC, and other major energy-consuming equipment. Emily will also identify potential energy-saving opportunities and recommend appropriate solutions.

496. Which of the following lighting technologies would Emily recommend to replace existing inefficient lighting fixtures for better energy efficiency and longer lifespan?
a. Incandescent bulbs
b. Compact fluorescent lamps (CFLs)
c. Light-emitting diode (LED) lamps
d. Halogen lamps


Answer: c. Light-emitting diode (LED) lamps
Explanation: LED lamps are the most energy-efficient lighting technology available. They have a significantly longer lifespan than other lighting options, reducing replacement costs and maintenance.

497. In addition to replacing inefficient lighting, what control system can Emily suggest to further improve lighting energy efficiency in the building?
a. Manual light switches
b. Timers
c. Occupancy sensors
d. Dimmer switches


Answer: c. Occupancy sensors
Explanation: Occupancy sensors can detect when a space is occupied and automatically turn lights on or off, reducing energy consumption by ensuring that lights are only on when needed.

498. To optimize the energy efficiency of the building's HVAC system, which of the following strategies should Emily consider?
a. Implementing a variable air volume (VAV) system
b. Installing additional heating and cooling equipment
c. Increasing the thermostat setpoints
d. Decreasing insulation levels


Answer: a. Implementing a variable air volume (VAV) system
Explanation: A variable air volume (VAV) system adjusts the amount of conditioned air delivered to different zones based on demand, improving energy efficiency by reducing the energy consumed for heating and cooling.

499. Which of the following devices can help reduce standby power consumption for office equipment in the building?
a. Uninterruptible power supply (UPS)
b. Advanced power strips
c. Ground fault circuit interrupter (GFCI)
d. Watt-hour meter

Answer: b. Advanced power strips
Explanation: Advanced power strips can reduce standby power consumption by automatically shutting off power to devices when they are not in use, such as during evenings and weekends.

500. What type of energy management system would Emily recommend to monitor and optimize energy usage across the building's various systems and equipment?
a. Building automation system (BAS)
b. Programmable logic controller (PLC)
c. Supervisory control and data acquisition (SCADA) system
d. Circuit breaker panel

Answer: a. Building automation system (BAS)
Explanation: A building automation system (BAS) can monitor and control various building systems and equipment, such as lighting, HVAC, and security, to optimize energy usage and reduce overall energy consumption.

Case Study
John is an electrical inspector responsible for ensuring the safety and compliance of new electrical installations in a residential building. He uses various inspection techniques, tools, and equipment to assess the installations and determine whether they comply with the relevant codes and standards. Upon completion of the inspection, John will issue a certification indicating that the installations meet the necessary requirements.

501. Before beginning the inspection, which safety equipment should John wear to protect himself from potential hazards?
a. Steel-toe boots
b. Personal flotation device
c. High-visibility vest
d. Insulated gloves

Answer: d. Insulated gloves
Explanation: Insulated gloves provide protection against electrical hazards such as shock and arc flash. They are essential for electrical inspectors when working near live electrical equipment.

502. Which of the following tests should John perform to verify the proper operation of ground fault circuit interrupter (GFCI) outlets in the residential building?
a. Continuity test
b. Insulation resistance test
c. GFCI functionality test
d. Earth loop impedance test

Answer: c. GFCI functionality test
Explanation: A GFCI functionality test checks whether the GFCI outlet trips correctly in response to a ground fault. This test ensures that the GFCI provides the required protection against electrical shock.

503. Which instrument would John use to measure the insulation resistance of electrical cables?
a. Ohmmeter
b. Multimeter
c. Megohmmeter
d. Ammeter

Answer: c. Megohmmeter
Explanation: A megohmmeter, also known as an insulation tester or a "Megger," is used to measure the insulation resistance of electrical cables, which is important for ensuring the safety and reliability of the electrical installation.

504. When inspecting electrical panelboards, which of the following is an important aspect for John to verify?
a. Proper labeling of circuit breakers
b. The presence of a main disconnecting means
c. Correct wire sizes for each circuit
d. All of the above

Answer: d. All of the above
Explanation: Proper labeling of circuit breakers, the presence of a main disconnecting means, and the correct wire sizes for each circuit are all essential aspects of a safe and compliant electrical panelboard installation.

505. Upon successful completion of the inspection, which type of document will John issue to certify that the electrical installations meet the required standards and codes?
a. Electrical permit
b. Notice of violation
c. Certificate of compliance
d. Work order

Answer: c. Certificate of compliance
Explanation: A certificate of compliance is issued by an electrical inspector to certify that an electrical installation meets the required standards and codes, indicating that the installation is safe and compliant.

*In conclusion, we have journeyed together through this comprehensive study guide, exploring the many facets of electrical installations, troubleshooting, and safety practices. We delved into residential and commercial wiring, circuit design, grounding and bonding, circuit protection devices, and many other crucial aspects of the electrical trade. Together, we discovered the importance of understanding codes and standards, energy efficiency, solar panel installations, and emergency and backup power systems. Throughout this journey, our aim was to provide you with the knowledge and confidence needed to excel in your electrical career or examination.*

*As you move forward, remember that you are not alone in facing challenges and overcoming obstacles. Your dreams are valid, and your pursuit of excellence is commendable. It is essential to learn from any setbacks or failures and use them as opportunities for growth and improvement. Fear not the unknown; instead, embrace the wealth of knowledge and experience that awaits you in the world of electrical work. Trust in your abilities and continue to seek guidance, knowledge, and support from those around you.*

*As you continue on your path, remember the wise words of Henry Ford: "Whether you think you can, or you think you can't – you're right." Believe in yourself and your potential, for it is through that belief that you will find success. As you face the challenges ahead, may you find strength, resilience, and inspiration in the knowledge you have gained from this study guide.*

*We wish you the very best in your endeavors, and may you achieve great success in your electrical career, exam, or project. Always aim high, and never lose sight of your dreams. Good luck, and may your journey be illuminated by the bright light of knowledge, skill, and determination.*

Printed in the USA
CPSIA information can be obtained
at www.ICGtesting.com
LVHW071752301223
767720LV00078B/2507